Soil Fertility Improvement and Integrated Nutrient Management - A Global Perspective

Soil Fertility Improvement and Integrated Nutrient Management - A Global Perspectivepics

Editor

Aleix Campo

scitus
academics

Soil Fertility Improvement and Integrated Nutrient Management - A Global Perspective
Edited by **Aleix Campo**

ISBN: 978-1-68117-583-6
Library of Congress Control Number: 2016946438

SCITUS Academics LLC,
616, Corporate Way, Suite 2, 4766,
Valley Cottage, NY 10989

www.scitusacademics.com

Notice

Printed in United States of America on Acid Free Paper

Preface

Crops need air, sun, water, and soil to thrive. When it comes to soil, however, quality usually trumps quantity. Rich and fertile land boasts a healthy mixture of phosphorous, potassium, and nitrogen, along with water, air, and soil micro-organisms that break down organic matter. Soil is an incredibly complex substance. It has physical and chemical properties that allow it to sustain living organisms—not just plant roots and earthworms, but hundreds of thousands of different insects, wormlike creatures and microorganisms. When these organisms are in balance, soil cycles nutrients efficiently, stores water and drains the excess, and maintains an environment in which plants can thrive. Plant nutrition is only one of more than fifty factors which directly affect both crop yield and quality. The availability of required nutrients, together with the degree of interaction between these nutrients and the soil, play a vital role in crop development. A deficiency in any one required nutrient or, a soil condition that limits or prevents a metabolic function from occurring can limit plant growth. A soil nutrient management plan should include analyzing soil deficiencies to determine the type, application rate, application interval, and the placement of any nutrients required to optimize short and long term productivity. Soil nutrient management involves not only the physical properties and mineral structure of the soil, but also the balance between soil pathogens and beneficial microbes. Beneficial microbes increase nutrient availability, reduce disease, reduce nutrient losses, and help degrade toxic compounds.

Soil Fertility Improvement and Integrated Nutrient Management - A Global Perspective provides a basic introduction to the biological, chemical, and physical properties affecting soil fertility and plant nutrition. The advances in the field of soil fertility are described in this book along with information regarding nutrient management.

Table of Contents

Chapter 1

Restoring Soil Quality to Mitigate Soil Degradation

Rattan Lal

The Ohio State University, Columbus, OH 43210, USA

ABSTRACT

Feeding the world population, 7.3 billion in 2015 and projected to increase to 9.5 billion by 2050, necessitates an increase in agricultural production of ~70% between 2005 and 2050. Soil degradation, characterized by decline in quality and decrease in ecosystem goods and services, is a major constraint to achieving the required increase in agricultural production. Soil is a non-renewable resource on human time scales with its vulnerability to degradation depending on complex interactions between processes, factors and causes occurring at a range of spatial and temporal scales. Among the major soil degradation processes are accelerated erosion, depletion of the soil organic carbon (SOC) pool and loss in biodiversity, loss of soil fertility and elemental imbalance, acidification and salinization. Soil degradation trends can be reversed by conversion to a restorative land use and adoption of recommended management practices. The strategy is to minimize soil erosion, create positive SOC and N budgets, enhance activity and species diversity of soil biota (micro, meso, and macro), and improve structural stability and pore geometry. Improving soil quality (*i.e.*, increasing SOC pool, improving soil structure, enhancing soil fertility) can reduce risks of soil degradation (physical, chemical, biological and ecological) while improving the environment. Increasing the SOC pool to above the critical level (10 to 15 g/kg) is essential to set-in-motion the restorative trends. Site-specific techniques of restoring soil quality include conservation agriculture, integrated nutrient management, continuous vegetative cover such as residue mulch and cover cropping, and controlled grazing at appropriate stocking rates. The strategy is to produce "more from less" by reducing losses and increasing soil, water, and nutrient use efficiency.

KEYWORDS

soil resilience; climate change; soil functions; desertification; soil carbon sequestration

1. INTRODUCTION

Of the 5.5 billion people living in developing countries in 2014 [1], a large proportion of them depend on agriculture for their livelihood. In fact, one billion of these people are small landholders who cultivate <2 ha of land [2]. With limited resources and poor access to inputs, management of soil quality is essential to strengthen and sustain ecosystem services. Soil degradation is a 21st century global problem that is especially severe in the tropics and sub-tropics. Some estimates indicate degradation decreased soil ecosystem services by 60% between 1950 and 2010 [3]. Accelerated soil degradation has reportedly affected as much as 500 million hectare (Mha) in the tropics [4], and globally 33% of earth's land surface is affected by some type of soil degradation [5]. In addition to negatively impacting agronomic production, soil degradation can also dampen economic growth, especially in countries where agriculture is the engine for economic development [6]. Over and above the environmental and economic impacts, there are also health risks of soil erosion [7] and other degradation processes [8].

Soil degradation implies a decline in soil quality [8] with an attendant reduction in ecosystem functions and services. Conceptually, there are four types of soil degradation: (i) physical; (ii) chemical; (iii) biological; and (iv) ecological (Figure 1). Soil physical degradation generally results in a reduction in structural attributes including pore geometry and continuity, thus aggravating a soil's susceptibility to crusting, compaction, reduced water infiltration, increased surface runoff, wind and water erosion, greater soil temperature fluctuations, and an increased propensity for desertification. Soil chemical degradation is characterized by acidification, salinization, nutrient depletion, reduced cation exchange capacity (CEC), increased Al or Mn toxicities, Ca or Mg deficiencies, leaching of NO_3-N or other essential plant nutrients, or contamination by industrial wastes or by-products. Soil biological degradation reflects depletion of the soil organic carbon (SOC) pool, loss in soil biodiversity, a reduction in soil C sink capacity, and increased greenhouse gas (GHG) emissions from soil into the atmosphere. One of the most severe consequences of soil biological degradation is that soil becomes a net source of GHG emissions (*i.e.*, CO_2 and CH_4) rather than a sink. Ecological degradation reflects a combination of other three, and leads to disruption in ecosystem functions such as elemental cycling, water infiltration and purification, perturbations of the hydrological cycle, and a decline in net biome productivity. The overall decline in soil quality, both by natural and anthropogenic factors, has strong positive feedbacks leading to a decline in ecosystem services and reduction in nature conservancy. Once the process of soil degradation is set-in-motion, often by land misuse and soil mismanagement along with the extractive farming, it feeds on itself in an ever-increasing downward spiral (Figure 2).

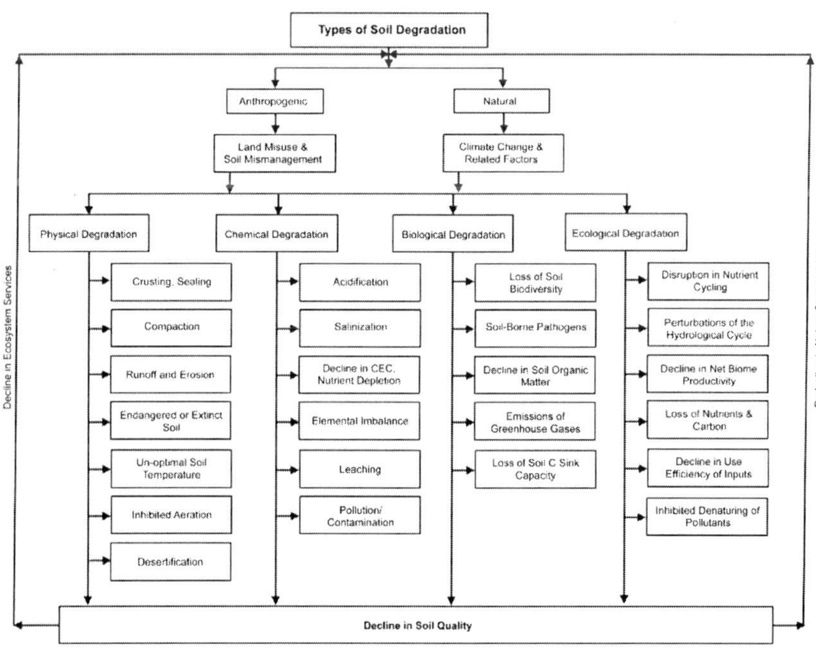

Figure 1. Types of soil degradation.

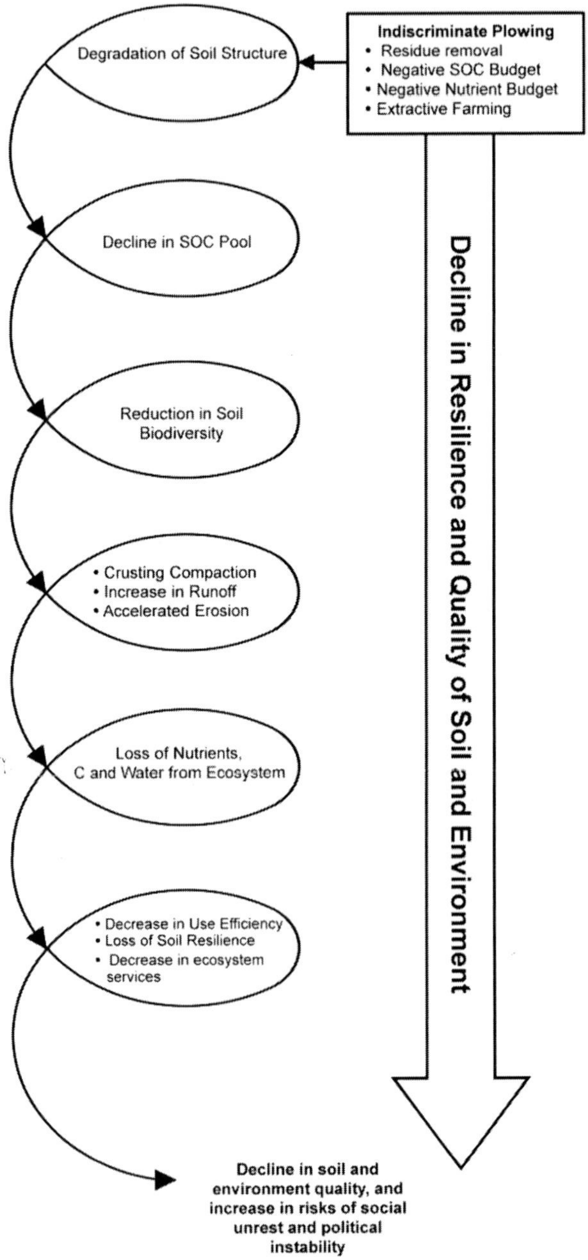

Figure 2. The downward spiral of decline in soil and environment quality exacerbated by indiscriminate plowing, residue removal and extractive farming.

The objectives of this review are to: (1) deliberate the role of soil resources in provisioning essential ecosystem services; (2) illustrate the impacts of soil degradation on decline in ecosystem services; and (3) identify strategies for improving soil quality to mitigate risks of soil degradation.

2. SOIL AND ECOSYSTEM SERVICES

Soil, the most basic of all resources, is the essence of all terrestrial life and a cultural heritage [9]. Yet, soil is finite in extent, prone to degradation by natural and anthropogenic factors, and is non-renewable over the human timescale (decades). Soil quality also has strong implications to human health [8,10], thus illustrating its important role in both society and the environment. Because of numerous ecosystem services provisioned through soils (e.g., food, feed, fiber, climate moderation through C cycling, waste disposal, water filtration and purification, elemental cycling) [11,12], soil quality must be protected or restored to enhance these services. Increased public awareness and a fundamental understanding of basic pedospheric processes (*i.e.*, biology, chemistry, physics, pedology, ecology) are essential both to enhancing long-term productivity and improving the environment [13].

3. SOIL ORGANIC CARBON AND ITS IMPACT ON SOIL QUALITY

The SOC pool, including its quantity and quality, is the defining constituent of soil [14,15]. Indeed, SOC pool is the most reliable indicator of monitoring soil degradation, especially that caused by accelerated erosion [16]. Soil degradation depletes the SOC pool, along with it, plant available N and other essential nutrients such as P and S. Furthermore, as identified repeatedly in this special issue of Sustainability, depletion of SOC pool is a global issue and a principal cause of soil degradation, especially in the European semi-arid Mediterranean regions [17]. Developing strategies to ensure the SOC pool is to increase and preferably maintain above the threshold or critical level of 10 to 15 g/kg (1.0%–1.5%), which is essential for reducing soil degradation risks and reversing degradation trends. Integrated nutrient management (INM) is one strategy that embodies sustainable management of the SOC pool and its dynamics [18]. Adoption of INM or similar management practices that create a positive soil/ecosystem C budget can not only increase productivity but also sequester additional atmospheric CO_2 into the SOC pool. This has been documented for many surface soils within the U.S. Corn Belt which act as C sinks when corn (*Zea mays* L.) is grown using recommended management practices (RMPs) such as conservation agriculture (CA) [19]. There also exists a strong relationship between vegetation cover and the SOC pool, such that excessive reductions in vegetation cover exacerbates risks of soil degradation and SOC depletion. A study conducted in the sub-tropical humid grasslands in South Africa indicated that the decline in grass (vegetative) cover from 100% to 0%–5% reduced the SOC pool by 1.25 kg/m^2 and the soil organic N (SON) pool by 0.074 kg/m^2 [20], There were also attendant declines in the C:N ratio and proportion of SOC and SON in the silt + clay fraction with the decline in aerial grass cover which negatively affected ecosystem functions of the acidic sandy loam soils. Similarly, transformation of a thicket vegetation to an open savanna (dominated by grasses) due to intensive grazing decreased soil quality in the Eastern Cape region of South Africa [21]. Indeed, savanna soils have lower SOC concentration and a greater tendency to crust than thicket soils because of the decreased quantity and stability of structural aggregates.

The widespread prevalence of degraded soils in sub-Saharan Africa (SSA), a classic example of a downward spiral, is attributed to over exploitation, extractive farming, low external inputs, and poor or improper management (Figure 2). Accelerated degradation is shrinking the finite soil resource even more rapidly in these regions of harsh climate and fragile soils. In this context, enhancing the SOC pool is important to sustain soil fertility and agronomic productivity [22]. Simply adding chemical fertilizers or improved varieties, as is often erroneously recommended even by well-intended advocates, is not enough.

The SOC pool of agricultural soils of West Africa, similar to those of croplands in other developing counties (e.g., South Asia), is severely depleted by over-exploitation of natural resources [23]. These soils must therefore be managed to increase both soil C and vegetation [24,25] to restore the degraded agroecosystem services. Changes in aerial vegetative cover could thus be used as an early indicator of shifts in soil ecosystem functions within fragile environments. A shift in vegetative cover may be caused by alterations in land use or climate change. In addition to SOC and SON pools, soil moisture regime is another important indicator of climate change [26]. In conjunction with changes in soil moisture regimes, projected global warming may also influence SOC decomposition rates [27,28] including that of fine woody debris. Field experiments have shown that warming increases mass loss for all vegetative species and size classes by as much as 30%. However, larger debris and that with higher initial lignin content decomposes more slowly than smaller debris and that with lower antecedent lignin content. Indeed, degradation of lignin may not follow the same trend as that of total mass loss [29]. Along with the adverse effects of soil erosion and other degradation processes, the SOC pool is also prone to climate change and associated alterations in temperature and moisture regimes.

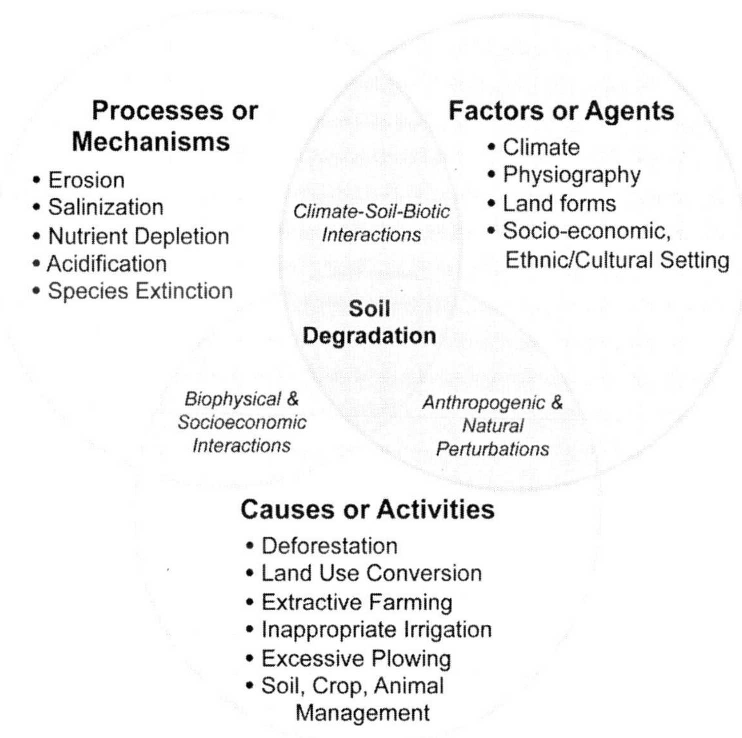

Figure 3. The process-factor-cause nexus as a driver of soil degradation.

The self-reinforcing soil degradation process (Figure 2) is strongly exacerbated by the interaction between processes, factors and causes of soil degradation (Figure 3). Processes include the mechanisms (types) of soil degradation. Factors comprise agents of degradation related to natural or anthropogenic drivers such as climate, physiography, socio-economic or ethnic/cultural parameters. Causes of soil degradation include specific activities which aggravate the adverse effects of processes and factors. Examples of specific causes include activities such as deforestation, land use conversion, extractive farming practices or over-exploitation, excessive grazing, excessive plowing *etc.* (Figure 3). The process-factor- cause nexus is strongly impacted by site-specific conditions. Thus, understanding the nexus or connectivity is critical to restoring soil quality and mitigating degradation.

4. SOIL QUALITY INDEX

The SOC pool is a key indicator of soil quality, and an important driver of agricultural sustainability. In addition to its amount, other parameters of SOC include its depth distribution, quality or attributes (physical, chemical, biological), and the turnover rate or the mean residence time (MRT). Relevant indicators of soil physical quality include amount and stability of aggregates; susceptibility to crusting and compaction; porosity comprising of pore geometry and continuity; water transmission (infiltration rate and amount) and retention as plant-available water capacity (PAWC); aeration and gaseous exchange; effective rooting depth; soil heat capacity and the temperature regime. Similarly, appropriate indicators of soil chemical quality include pH, CEC, nutrient availability; and favorable elemental balance and lack of any toxicity or deficiency. Relevant indicators of soil biological quality are microbial biomass C (MBC), activity and diversity of soil fauna and flora, absence of pathogens and pests as indicated by a soil's disease-suppressive attributes. An optimal combination of these properties affects agronomic productivity; use efficiency of water, nutrients and other inputs; and sustainability of management systems. Indicators of soil quality differ among soil types, climates and land uses. For example, there are specific soil quality indicators for the intensively managed soils of the Indo-Gangetic Plains [30] that will differ from those for tropical Alfisols in semi-arid regions [31,32,33,34]. A spectral soil quality index based on application of reflectance spectroscopy has also been proposed as a diagnostic tool to assess soil quality [35]. This technique can provide a characterization of physical, chemical and biological attributes that can be merged together to indicate how well a soil is functioning for a specific use [36,37,38,39].

5. CONSERVATION AGRICULTURE AND SOIL QUALITY

Four basic principles of CA are [40]: (i) retention of crop residue mulch; (ii) incorporation of a cover crop in the rotation cycle; (iii) use of INM involving combination of chemical and bio fertilizers; and (iv) elimination of soil mechanical disturbances. Properly implemented on suitable soil types, CA has numerous co-benefits including reduced fuel consumption and increased soil C sequestration. Mechanical tillage is an energy-intensive process [41] and its reduction or elimination can decrease consumption of fossil fuels. For example, conversion from plow tillage (PT) to CA can reduce diesel consumption by as much as 41 L/ha [42].

In addition, an increase in SOC pool under CA can occur in soils not prone to accelerated erosion, and those which have optimal management strategies. A modeling study in Western Kenya showed that site-specific optimal management strategies can lead to SOC pool of 20 to 40 Mg/ha in 0.1 m depth and corn grain yield of 3.5 to 4.2 Mg/ha [43]. The most desirable tillage systems are those which restore soil quality, minimize risks of soil erosion, improve use efficiency of rain water and fertilizers [44] and minimize risks of SOC and nutrient depletion. Impacts of CA on soil quality restoration, an example of an upward spiral, is outlined in Figure 4.

Indiscriminate use of plowing, coupled with excessive removal of crop residues and unbalanced use of chemical fertilizers, can degrade soil quality, deplete SOC pool, and aggravate risks of soil erosion (Figure 2). In contrast, conversion of plow/traditional tillage to CA, especially on sloping lands and those vulnerable to accelerated erosion by water and wind under conventional management, can be conservation-effective, reverse degradation trends, and set-in-motion soil restoration processes with an upward spiral (Figure 4). Retention of crop residue mulch, and incorporation of a cover crop (forages) in the rotation cycle while eliminating bare fallows, can conserve soil and water and improve SOC pool in the surface layer (Figure 4). Increases in soil biodiversity, MBC and activity of earthworms and termites can all improve aggregation and encapsulate C within stable micro-aggregates as outlined in the hierarchy concept [47]. Strengthening elemental cycling, in conjunction with coupled cycling of C and H_2O, can increase the solum's C sink capacity and soil profile depth through increased bioturbation by earthworms or termites, and use of deep-rooted plants such as pigeon pea (*Cajanus cajan*), townsville stylo (*Stylosanthes humilis* (Kunth) Hester), or alfalfa (*Medicago sativa*). Long-term (>10 years) soil quality improvement will increase net biome productivity, improve water and nutrient use efficiencies, and increase above and below-ground biomass-C within the ecosystem. Progressive improvements in rhizospheric processes, driven by biotic mechanisms, would restore soil quality and mitigate degradation (Figure 4). It is important to note, however, that CA is a holistic and system-based approach. Mere elimination of plowing, while removing an excessive amount of crop residues and biomass for other uses (e.g., biofuels, industrial purposes) is not CA, and is rather an extractive farming system with negative impacts on soil and the environment. Thus, comparative analyses of un-related datasets can lead to erroneous assessments of SOC sink capacity associated with a properly implemented CA system [45] and misinterpretation of agronomic yields [46]. Furthermore, improvements in soil quality require more than the input of new varieties and chemical fertilizers [48]. The real issue is improving physical, biological and ecological components of soil quality, and set-in-motion an upward spiral eventually leading to social and political stability and international security (Figure 4).

6. SOIL FERTILITY MANAGEMENT TO RESTORE SOIL QUALITY

Sustainable intensification (SI), producing more from less by reducing losses and increasing the use efficiency, is attainable only through improvement of soil quality including chemical quality or soil fertility. Although not the only way to increase soil fertility, the use of INM is a very effective approach for achieving SI. Nutrient depletion and loss of soil fertility are major causes of low productivity [49] in many developing countries. Use of organic amendments, by recycling organic by-products including urban waste, is a useful strategy to enhance soil fertility [50], and improve structural stability or aggregates [51]. While, nitrogen (N) input is important to improving soil fertility, its improper and/or excessive use can also lead to environmental pollution. China consumes about 30% of the world's N fertilizer [52], and is able to feed ~22% of the world population on just 6.8% of the global cropland area. However, the country has severe environmental problems because of low N use efficiency, leaching of reactive N into surface and groundwater resources, and emission of N (as N_2O) into the atmosphere.

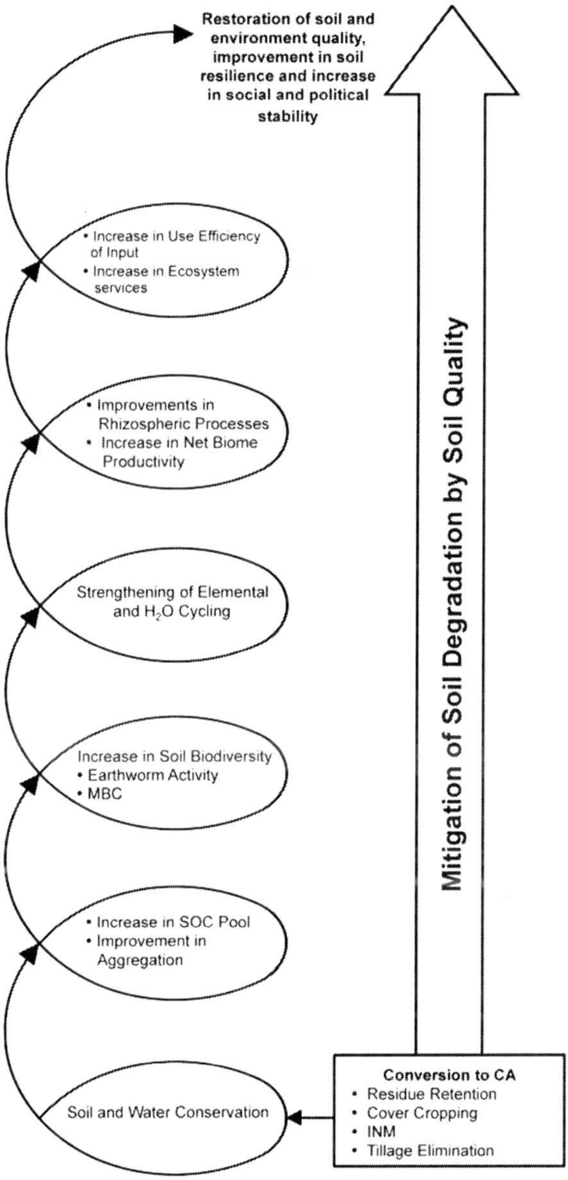

Figure 4. Increase in soil resilience and mitigation of soil degradation by conservation agriculture (CA). Meta-analyses and any other comparisons among unrelated soil management practices can lead to misinterpretation of SOC sink capacity by CA [45] and erroneous inferences on agronomic productivity [46]. The mission is to identify site-specific packages of CA practices to make it functional (INM = integrated nutrient management).

7. SOIL QUALITY AND WATER RESOURCES

High soil quality or a healthy soil provides the foundation for a healthy economy, environment, and terrestrial biosphere. Thus, there exists a close link between soil quality and water resources in close proximity, such as the health of coastal ecosystems [53]. Changes in land use often affect water quality and pollutant loading [54,55]. Off-site movement of agricultural chemicals is often a significant source of non-point pollution. Many of the major rivers in countries with emerging economies have severe water pollution, contamination and eutrophication problems [56]. Downstream areas are often adversely affected because of adhoc agricultural development activities upstream. Among the most important adverse impacts are river desiccation, ground water depletion, surface and ground water pollution, accelerated erosion, sedimentation, salinization, and nutrient depletion [57]. These problems are especially severe in densely populated regions (e.g., East Asia, South Asia). Rapid urbanization, industrialization and increases in water demand have created severe water quality problems and degradation of the Ganges [58]. Land use changes induced by urbanization have had strong impacts both on soil and water quality in northern Iran [54]. Release of P from agricultural sources to streams and rivers has become a major factor affecting surface water quality not only in China [59], but also in the Great Lakes of North America (e.g., algal bloom of 2014). In Thailand, agriculture [especially rice (*Oryza sativa*) farming] is the key nutrient source within the Thachin River Basin [56]. Recent industrialization and community development have exacerbated water pollution and habitat degradation in the Gulf of Thailand [60]. The problem has been exacerbated by a rapid decrease in mangrove forest, coral reefs, and fishery resources due to misuse and mismanagement.

Irrigated agriculture, an important management strategy for high agronomic productivity in arid and semi-arid regions, is a mixed blessing. Mismanagement of irrigation waters has exacerbated problems with saline-sodic soils which now occupy more than 20% of the irrigated lands [61]. Furthermore, arid wetlands are also prone to contamination by sub-surface agriculture irrigation and drainage as in the Western USA [62]. These areas often experience toxicity problems in fish and wildlife due to drainage of water contaminants. In response, provisions must be made to reduce the amount of contaminants entering wetlands, and to provide for better allocation of freshwater between agriculture and wildlife [62]. Salinity problems are also often confounded by the reuse of untreated waste water (gray/black water) in agriculture [63], especially in urban areas prone to water shortages or those having water resources of marginal quality.

Restoring soil quality within managed ecosystems is critical to improving and sustaining water quality. To accomplish this goal, it is essential to develop strategies for integrated management of soil and water resources because of their strong inter-connectivity or the soil-water-waste nexus. While integrated water management alone is useful [64], the importance of the soil-water nexus cannot be over-emphasized [24]. Management of sediments, especially contaminated ones [65], is another important component of the soil-water nexus that must be critically examined.

8. STRATEGIES FOR SOIL QUALITY RESTORATION

Restoring the quality of degraded soils is a challenging task, especially in regions dominated by small, resource-poor landholders. Re-carbonization of the depleted SOC pool, which is essential to numerous functions, requires regular input of biomass-C and essential elements (*i.e.*, N, P, and S) [66]. Thus, restoration of soil quality is a societal, national and international task that necessitates a coordinated approach.

There are three basic strategies of restoring soil quality (Figure 5): (i) minimizing losses from the pedosphere or soil solum; (ii) creating a positive soil C budget, while enhancing biodiversity; and (iii) strengthening water and elemental cycling. There is no silver bullet or panacea to accomplish these basic tasks, and site-specific factors (biophysical, social, economic, cultural) play a significant role. Some examples of site-specific, RMPs are outlined in Table 1. However, each of these technologies has their own tradeoffs, which must be duly considered and minimized.

8.1. Soil Erosion Management

Soil erosion must be curtailed to within the tolerable limits, which is often much less than the presumed value of 12.5 Mg/ha per year. Accelerated erosion also depletes the SOC pool and nutrient reserves. In general, the enrichment ratio of SOC, clay and essential plant nutrients (N, P, S) is >1 (and most often as much as 5 or more) because of the preferential removal of these constituents. Conversion from PT to CA can reduce risks associated with soil erosion and nutrient loss while also providing numerous on- and off-site benefits (Figure 4) [12]. An important strategy is to establish cause-effect relationships, alleviate the causative factors and minimize the risks. Accelerated erosion is a symptom of land misuse and soil mismanagement. Reductions in plant cover caused by over-grazing and the trampling effect can degrade soil structure, reduce water infiltration, increase runoff, aggravate soil erosion, and cause severe economic losses [76]. As an example, experiments conducted in South Africa indicated that plant cover reduction by overgrazing significantly decreased the SOC pool with strong impact on the C cycle [77]. In arid regions, fire-induced depletions of the vegetation cover can also exacerbate the problem, especially after a torrential rainfall because ash left on the soil surface can aggravate hydrophobicity by creating an obtuse contact angle between the solid and liquid phases. When the protective litter cover is burned, the very first rainfall generally results in high surface runoff and aggravates erosion as was the classic situation of wild fire in May–June 2000 in hills which threatened the Los Alamos National Laboratory by runoff and sedimentation. Experiments conducted in Spain indicated high post-fire soil degradation risks and the need for identification of a short-term strategy to conserve soil and water on steep slopes with erodible soils [78]. In addition to the adverse effects on soil quality and productivity, there are also health risks associated with soil erosion [79]. This is especially true for regions prone to wind erosion and dust storms, such as the Harmattan in the Sahel [80].

8.2. Improving Soil/Agro-Biodiversity

Soil biota are important to soil quality and reduce risks of degradation and desertification. Indeed, soil biota comprise a major component of global terrestrial biodiversity and perform critical roles in key ecosystem functions (e.g., biomass decomposition, nutrient cycling, moderating CO_2 in the atmosphere, creating disease suppressive soils, *etc.*). Improving activity and species diversity of soil fauna and flora (micro, meso and macro) is therefore essential to restoring and improving soil quality and reducing risks of soil degradation. Adverse effects of agricultural management on soil microbiological quality is another global concern. As a management tool, either a microbiological quality index [81] or a microbiological degradation index [82] can be useful for decision-making processes [82] Relevant parameters include MBC, respiration, water soluble carbohydrates, enzymatic activities, dehydrogenase activity and activities of other important hydrolases (e.g., urease, protease, phosphatas and β-glucosidase) [82]. There are also marked seasonal changes in biotic and abiotic factors that affect the biological component of soil resources. Vegetative cover, influenced by seasonal changes, has a strong impact on soil microbiological processes. In degraded soils of arid and semi-arid regions, changes in soil moisture regimes can also affect MBC and activity [83].

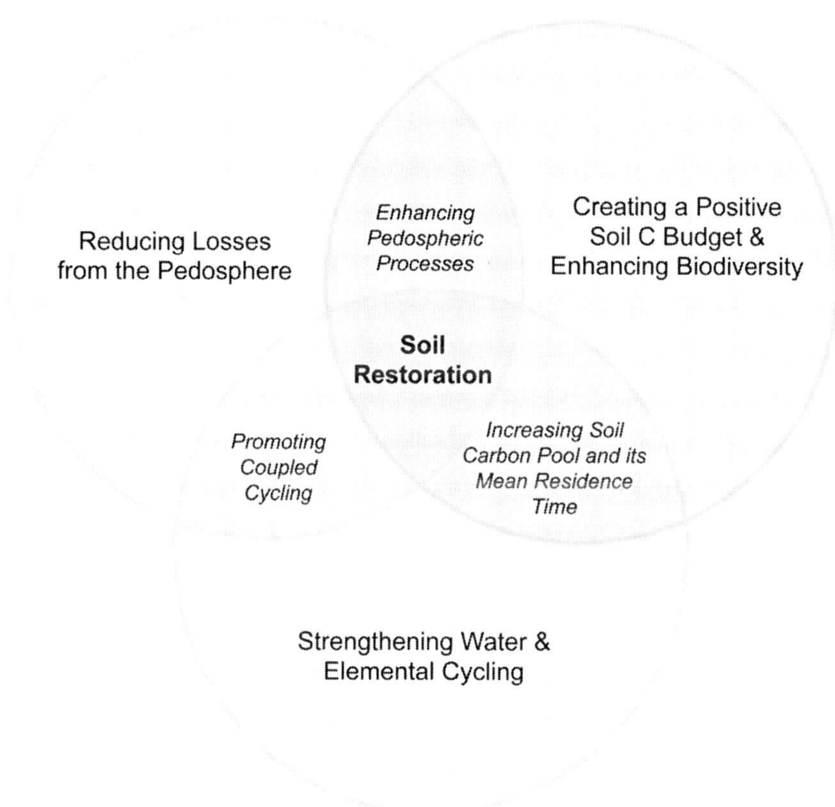

Figure 5. Three strategies of restoring and managing soil quality for mitigating risks of soil degradation.

Table 1. Strategies to improve soil quality.

Strategy	Region	Process	Reference
Litter turnover	Tropics	The rate of organic matter and C supply and nutrient cycling reactivation	[3]
Forestry Plantations	Tropics	Silvo-pastoral system for nutrient cycling	[67]
Woodlot Islets	Degraded drylands	Silvo-pastoral systems in drylands	[68]
Soil Carbon Sequestration	Agroecosystems	Optimal management strategies	[69]
Integrated Nutrient Management	Sub-Saharan Africa	Soil quality management	[17]
Nutrient Management for SOC Sequestration	Sub-Tropical Red Soils (China)	Soil carbon buildup	[70]
Manuring	Indus Plains	Application of farm manure	[71]
Residue Retention as Mulch	Mexican Highlands	Improvement of soil structure	[72]
Regular Organic Inputs	Western Kenya	Nutrient retention and soil structure improvement	[43,73]
Urban Waste	Mediterranean Europe	Enhancing soil fertility	[16,74]
Soil Biological Management	Global soils	Enhance ecosystem services provisioned by SOC pool	[15]
Environmental Awareness	U.S.	Promoting technology adoption	[75]

The importance of macro-organisms (e.g., earthworms, termites) for restoring soil quality has been widely recognized for centuries [84]. Conversion of PT to CA, with crop residue mulch and cover cropping (Figure 4), can increase earthworm activity and also improve structural properties [85,86], but the conversion can also have implications regarding transport of agricultural pollutants into the drainage water [85]. Experiments conducted in central Mexico indicated that conversion of PT to CA improved soil surface aggregation and aggregate stability, increased water infiltration, and enhanced most parameters related to soil quality [87]. Therefore, risks of soil degradation can be mitigated through adoption of land use and management systems which improve soil biological processes, and introduction of beneficial organisms into soils by selective inoculation. For these and other reasons, the presence of earthworms, termites and other soil biota are often identified as important indicators of quality in tropical soils [86,88].

8.3. Soil Restorative Farming/Cropping Systems

Farming/cropping systems (rotations, soil fertility management, erosion control, grazing/stocking rate, water management) affect the type, rate and severity of soil degradation by altering the SOC pool, structural morphology, and other properties. Specifically, crop rotations and grazing can significantly impact SOC pool and the attendant soil properties [89]. Similar to arable lands, managing quality of rangeland soils is also essential for reducing risks of degradation. Sustainable management of rangeland soils is especially challenging because of high variability, harsh environments, and the temptation for over-grazing. A reduction in the proportion of palatable perennials, increases in densification (compaction), and declines in SOC are some of the constraints that need to be alleviated [90,91,92]. An important strategy to reduce the risks of degrading rangeland soils is to conserve and efficiently manage soil water through an improved understanding of the hydrological attributes [93,94]. Under West African conditions, construction of stone bunds and establishment of contour vegetative hedgerows can be effective for water conservation [95]. Establishment and management of forage trees (*i.e.*, *Acacia fadherbia*) [96] and grass-legume mixtures [97] can also improve the quality of rangeland soils.

9. SOIL RESILIENCE

The term soil resilience refers to the ability of the soil to recover its quality in response to any natural or anthropogenic perturbations. Soil resilience is not the same as soil resistance, because resilience refers to "elastic" attributes that enable a soil to regain its quality upon alleviation of any perturbation or destabilizing influence [98,99]. Sound rhizospheric processes are essential for soil resilience against anthropogenic/natural perturbations. Being a dominant site of microbial metabolism, it is pertinent to identify management systems that stimulate soil microbiotic activity and related microbial processes. In this context, an "eco-physiological index" has been proposed to assess the impact of soil resilience [100] on soil processes. Managing the quantity and quality of SOC pool is once again a crucial guiding principle in identifying appropriate management practices that will strengthen resilience and reduce risks for soil degradation [101]. The SOC pool size is strongly related to the quantity of both above and below-ground biomass-C inputs. It is the assured, continuous input of the biomass-C that moderates MBC, provides a reservoir of plant nutrients (e.g., N, P, S), influences nutrient cycling, and improves/stabilizes soil structural morphology and geometry [98,99]. The so-called "sustainable land management (SLM)" concept is based on similar strategies of preserving productivity of the resource base for future generations [101]. There are also some organic management options for reducing soil degradation risk and improving human health [102], that may have site-specific niches. Biochar, a C-rich soil amendment derived from biomass by pyrolysis, can be produced from human sewage [103] and used to improve soil resilience [98] while also mitigating climate change.

In addition to biotic techniques, ancient farmers also developed mechanical/engineering techniques to sustain and improve their soils. Terraced agriculture evolved independently at several locations around the world (e.g., East Asia, the Himalayan region, Yemen, the Andean region). A study of pre-Columbian terraces from the Paca Valley, Peru, indicated that soil depletion from cultivation has compromised soil quality through loss of fine material and SOC. Overall, however, Paca Valley terraces have improved topsoil retention and supported deep profile with a good soil resilience [104].

There are no universally applicable techniques of managing soil resilience, but there are several approaches for ensuring sustainable soil management. Each of these approaches has tradeoffs that must be objectively and critically assessed (Table 1, Figure 6). In view of the heavy demands for agricultural produce to meet the needs of the growing and increasingly affluent population and emerging economies (e.g., India, China, Brazil, Mexico), the role of agricultural practices and their impact on soil, climate, gaseous emission, water resources, biodiversity, along with economic, political, social and ecologic dimensions [105] must be considered more now than in the past. The ideal strategy is to meet increasing global food demands while simultaneously restoring soil quality, improving the environment, and minimizing the tradeoffs.

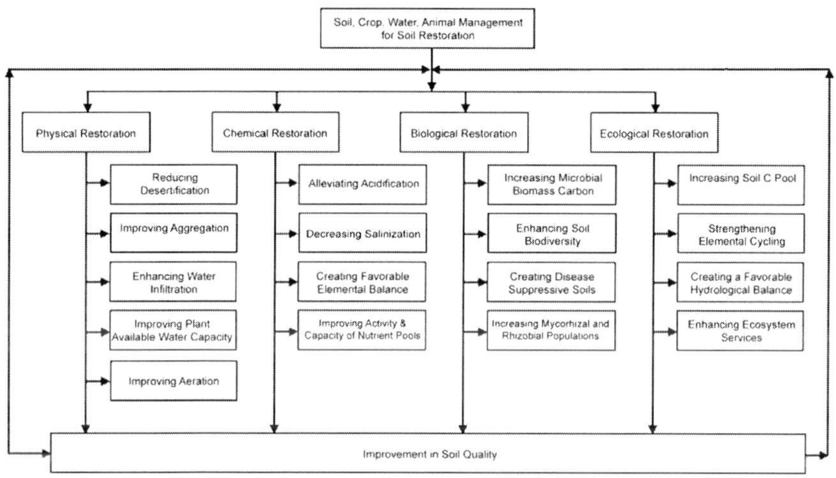

Figure 6. Strategies of restoring soil quality.

In the context of social, economic and cultural issues, it would be a serious omission to ignore poverty, human drudgery and social/gender equity. There is a strong link between poverty and soil quality. When people are poor, desperate and hungry, they pass on their sufferings to the land. Yet, some sociologists have questioned poverty as a major cause of soil/environmental degradation [106]. Stewardship and desperateness are mutually exclusive. Yield gaps, due to lack of adoption of RMPs in SSA and SA, are poverty traps that require a paradigm shift [107]. Since the 1980s, China has bridged the yield gap, alleviated poverty and improved soil quality [108]. Nonetheless, environmental issues remain to be addressed, effectively and immediately, to make China's agricultural revolution a successful venture [109].

10. PEAK SOIL *VS.* ENDANGERED SOIL

Endangered soil [110] and peak soil (losing soil more quickly than it is replaced [111] are concepts that need to be considered philosophically and scientifically because soil resources are finite and non-renewable over the human time frame. They are also geographically disturbed in a non-uniform manner. Excessive plowing, accelerated erosion, and over fertilization are all depleting soil resources, threatening food security, and jeopardizing the environment. Food insecurity is made even more acute by scarcity of soil resources of good quality (prime land) and risks of soil degradation [112]. Soil degradation is affecting 33% of all soils. Indeed, soil is an essentially forgotten resource [113]. In response, a holistic management approach is needed to improve soil quality. Site-specific and appropriate land husbandry practices must be identified to restore physical, chemical, biological, and ecological components of soil quality [114]. The goal should be to increase productivity per unit area, time, and energy input; while restoring soil quality and reducing environmental degradation risks. In terms of soil quality, ecosystem services provisioned by the SOC pool cannot be by-passed by applying commercial chemicals or other technologies [15]. Only by increasing SOC pool can the need for additional inorganic fertilizer N be reduced, especially in degraded soils. Conversion to CA, combined with residue retention and other components, is a sustainable option for several soil-specific conditions (Figure 4). Environmental awareness and stewardship are also important for improving adoption of RMPs and promoting soil restoration.

11. CONCLUSIONS

Soil resources are finite in extent, unequally distributed geographically, prone to degradation by land misuse and mismanagement, but essential to all terrestrial life and human wellbeing. Soil degradation can be physical (e.g., decline in structure, crusting, compaction, erosion, anaerobiosis, water imbalance), chemical (e.g., acidification, salinization, elemental imbalance comprising of toxicity or deficiency, nutrient deficiency), biological (depletion of SOC pool, reduction in soil biodiversity, decline in microbial biomass-C), or ecological (e.g., disruption in elemental cycling, decline in C sink capacity). Soil degradation leads to reduction in ecosystem functions and services of interest to human and conservation of nature. The SOC pool, its amount and depth-distribution along with turnover and mean residence time, is a critical component of soil quality and source of numerous ecosystem services. Soil degradation depletes SOC pool, and its restoration to threshold levels of at least 11 to 15 g kg^{-1} (1.1%–1.5% by weight) within the root zone is critical to reducing soil and environmental degradation risks. Important strategies for soil quality restoration and reducing environmental degradation risks are: (i) reducing soil erosion; (ii) creating a positive soil/ecosystem C budget; (iii) improving availability of macro (N, P, S) and micro-nutrients (Zn, Fe, Cu, Mo, Se); (iv) increasing soil biodiversity especially the microbial process; and (v) enhancing rhizospheric processes. The ultimate goal should be to adopt a holistic and integrated approach to soil resource management. The finite nature of soil resources must never be taken for granted—they must be used, improved, and restored.

REFERENCES

1. Van Pham, L.; Smith, C. Drivers of agricultural sustainability in developing countries: A review. *Environ. Syst. Decis.*__2014__, *34*, 326–34.

2. IFAD. *The Rural Poverty Report 2011*; International Fund for Agricultural Development: Rome, Italy, 2010.

3. Leon, J.; Osorio, N. Role of Litter Turnover in Soil Quality in Tropical Degraded Lands of Colombia. *Sci. World J.* **2014**,*13*.

4. Lamb, D.; Erskine, P.; Parrotta, J. Restoration of degraded tropical forest landscapes. *Science* **2005**, *310*, 1628–1632.

5. Bini, C. Soil: A precious natural resource. In *Conservation of Natural Resources*; Kudrow, N.J., Ed.; Nova Science Publishers: Hauppauge, NY, USA, 2009; pp. 1–48.

6. Scherr, S.J. The future food security and economic consequences of soil degradation in the developing world. In*Response to Land Degradation*; Oxford Press: New Delhi, India, 2001; pp. 155–170.

7. Guerra, A.; Marcal, M.; Polivanov, H.; Lima, N.; Souza, U.; Feitosa, A.; Davies, K.; Fullen, M.A.; Booth, C.A. Environment management and health risks of soil erosion gullies in São Luíz (Brazil) and their potential remediation using palm-leaf geotextiles. In *Environmental Health Risk II*; WIT Press: Southampton, UK, 2005; pp. 459–467.

8. Lal, R. Soil degradation as a reason for inadequate human nutrition. *Food Sec.* **2009**, *1*, 45–57.

9. Bini, C.; Zilioli, D. Is Soil a Cultural Heritage? Proc. III Conv. Int. Architettura del Passaggio, 20 suppl. DVD. Available online: http://arca.unive.it/handle/10278/22330 (accessed on 12 May 2015).

10. Abrahams, P. Soils: Their implications to human health. *Sci. Total Environ.* **2002**, *291*, 1–32.

11. Robinson, D.; Emmett, B.; Reynolds, B.; Rowe, E.; Spurgeon, D.; Keith, A.; Lebron, I.; Hockley, N.; Hester, R.; Harrison, R. Soil Natural Capital and Ecosystem Service Delivery in a World of Global Soil Change. *Soils Food Secur.***2012**, *35*, 41–68.

12. Graber, D.R.; Jones, W.A.; Johnson, J.A. Human and ecosystem health. *J. Agromedicine* **1995**, *2*, 47–64.

13. Singer, M.J.; Warkentin, B.P. Soil in an environmental context: An American perspective. *Catena* **1996**, *27*, 179–189.

14. Krupenikov, I.; Boincean, B.; Dent, D.; Krupenikov, I.; Boincean, B.; Dent, D. Humus-Guardian of Fertility and Global Carbon Sink. In *The Black Earth: Ecological Principles for Sustainable Agriculture on Chernozem Soils*; International Year of Planet Earth 2011; Springer: Dordrecht, The Netherlands, 2011; pp. 39–50.

15. Manlay, R.; Feller, C.; Swift, M. Historical evolution of soil organic matter concepts and their relationships with the fertility and sustainability of cropping systems. *Agric. Ecosyst. Environ.* **2007**, *119*, 217–233.

16. Rajan, K.; Natarajan, A.; Kumar, K.; Badrinath, M.; Gowda, R. Soil organic carbon—the most reliable indicator for monitoring land degradation by soil erosion. *Curr. Sci.* **2010**, *99*, 823–827.

17. Diacono, M.; Montemurro, F. Long-term effects of organic amendments on soil fertility: A review. *Agron. Sustain. Dev.***2010**, *30*, 401–422.

18. Vanlauwe, B.; Hester, R.; Harrison, R. Organic Matter Availability and Management in the Context of Integrated Soil Fertility Management in sub-Saharan Africa. *Soils Food Secur.* **2012**, *35*, 135–157.

19. Clay, D.; Chang, J.; Clay, S.; Stone, J.; Gelderman, R.; Carlson, G.; Reitsma, K.; Jones, M.; Janssen, L.; Schumacher, T. Corn Yields and No-Tillage Affects Carbon Sequestration and Carbon Footprints. *Agron. J.* **2012**, *104*, 763–770.

20. Dlamini, P.; Chivenge, P.; Manson, A.; Chaplot, V. Land degradation impact on soil organic carbon and nitrogen stocks of sub-tropical humid grasslands in South Africa. *Geoderma* **2014**, *235*, 372–381

21. Mills, A.; Fey, M. Transformation of thicket to savanna reduces soil quality in the Eastern Cape, South Africa. *Plant Soil* **2004**, *265*, 153–163.

22. Tiessen, H.; Cuevas, E.; Chacon, P. The role of soil organic-matter in sustaining soil fertility. *Nature* **1994**, *371*, 783–785.

23. Lal, R. Soil carbon sequestration impacts on global climate change and food security. *Science* **2004**, *304*, 1623–1627.

24. Lal, R. Soil carbon sequestration in natural and managed tropical forest ecosystems. *J. Sust. For.* **2005**, *21*, 1–30.

25. Batjes, N. Options for increasing carbon sequestration in West African soils: An exploratory study with special focus on Senegal. *Land Degrad. Dev.* **2001**, *12*, 131–142.

26. Eaton, W.; Roed, M.; Chassot, O.; Barry, D. Differences in soil moisture, nutrients and the microbial community between forests on the upper Pacific and Caribbean slopes at Monteverde, Cordillera de Tilaran: Implications for responses to climate change. *Trop. Ecol.* **2012**, *53*, 235–240.

27. Melillo, J.M.; Steudler, P.A.; Tian, H.; Butler, S. Fertilizing change: Carbon-nitrogen interactions and carbon storage in land ecosystems. In *Handbook of Climate Change and Agroecosystems: Impact, Adaptation*

and Mitigation; Hillel, D., Rosenzweig, C., Eds.; Imperial College Press: London, UK, 2010; pp. 21–36.

28. Melillo, J.; Steudler, P.; Aber, J.; Newkirk, K.; Lux, H.; Bowles, F.; Catricala, C.; Magill, A.; Ahrens, T.; Morrisseau, S. Soil warming and carbon-cycle feedbacks to the climate system. *Science* **2002**, *298*, 2173–2176.

29. Berbeco, M.; Melillo, J.; Orians, C. Soil warming accelerates decomposition of fine woody debris. *Plant Soil* **2012**, *356*, 405–417.

30. Ray, S.; Bhattacharyya, T.; Reddy, K.; Pal, D.; Chandran, P.; Tiwary, P.; Mandal, D.; Mandal, C.; Prasad, J.; Sarkar, D.;*et al*. Soil and land quality indicators of the Indo-Gangetic Plains of India. *Curr. Sci.* **2014**, *107*, 1470–1486.

31. Sharma, K.; Grace, J.; Mandal, U.; Gajbhiye, P.; Srinivas, K.; Korwar, G.; Bindu, V.; Ramesh, V.; Ramachandran, K.; Yadav, S. Evaluation of long-term soil management practices using key indicators and soil quality indices in a semi-arid tropical Alfisol. *Aust. J. Soil Res.* **2008**, *46*, 368–377.

32. Karlen, D.; Mausbach, M.; Doran, J.; Cline, R.; Harris, R.; Schuman, G. Soil quality: A concept, definition, and framework for evaluation. *Soil Sci. Soc. Am. J.* **1997**, *61*, 4–10.

33. Karlen, D.; Stott, D.; Doran, J.; Coleman, D.; Bezdicek, D.; Stewart, B. A framework for evaluating physical and chemical indicators of soil quality. *Defin. Soil Q. Sustain. Environ.* **1994**, *35*, 53–72.

34. Andrews, S.S.; Carroll, C.R. Designing a soil quality assessment tool for sustainable agroecosystems management.*Ecol. Appl.* **2001**, *11*, 1573–1585.

35. Paz-Kagan, T.; Shachak, M.; Zaady, E.; Karnieli, A. A spectral soil quality index (SSQI) for characterizing soil function in areas of changed land use. *Geoderma* **2014**, *230*, 171–184.

36. Lal, R. *Methods and Guidelines for Assessing Sustainable Use of Soil and Water Resources in the Tropics*; USDA/SMSS Bull: 21; US Department of Agriculture: Washington, DC, USA, 1994.

37. Andrews, S.; Karlen, D.; Cambardella, C. The soil management assessment framework: A quantitative soil quality evaluation method. *Soil Sci. Soc. Am. J.* **2004**, *68*, 1945–1962.

38. Andrews, S.; Karlen, D.; Mitchell, J. A comparison of soil quality indexing methods for vegetable production systems in Northern California. *Agric. Ecosyst. Environ.* **2002**, *90*, 25–45.

39. Gugino, B.K. *Cornell Soil Health Assessment Training Manual*; New York State Agricultural Experiment Station Cornell University: New York, NY, USA, 2009.

40. Lal, R. On Sequestering Carbon and Increasing Productivity by Conservation Agriculture. *J. Soil Water Conserv.* **2015**. in press.

41. Rodríguez, Y.E.O.; Hernández, D.C. Energetic balance of two farming tools in a Fluvisol for the cultivation of sweet potato (Ipomoea batatas Lam). *Rev. Cienceis Tecinas Agropecu.* **2013**, *22*, 21–25.

42. Labreuche, J.; Lellahi, A.; Malaval, C.; Germon, J. Impact of no-tillage agricultural methods on the energy balance and the greenhouse gas balance of cropping systems. *Cah. Agric.* **2011**, *20*, 204–215.

43. Kimetu, J.; Lehmann, J.; Ngoze, S.; Mugendi, D.; Kinyangi, J.; Riha, S.; Verchot, L.; Recha, J.; Pell, A. Reversibility of soil productivity decline with organic matter of differing quality along a degradation gradient. *Ecosystems* **2008**, *11*, 726–739.

44. So, H.; Kirchhof, G.; Bakker, R.; Smith, G. Low input tillage/cropping systems for limited resource areas. *Soil Tillage Res.* **2001**, *61*, 109–123.

45. Powlson, D.; Stirling, C.; Jat, M.; Gerard, B.; Palm, C.; Sanchez, P.; Cassman, K. Limited potential of no-till agriculture for climate change mitigation. *Nat. Clim. Chang.* **2014**, *4*, 678–683.

46. Pittelkow, C.; Liang, X.; Linquist, B.; van Groenigen, K.; Lee, J.; Lundy, M.; van Gestel, N.; Six, J.; Venterea, R.; van Kessel, C. Productivity limits and potentials of the principles of conservation agriculture. *Nature* **2015**, *517*, U365–U482.

47. Tisdall, J.; Oades, J. Organic-matter and water-stable aggregates in soils. *J. Soil Sci.* **1982**, *33*, 141–163.

48. Sanchez, P.A. En route to plentiful food production in Africa. *Nat. Plants* **2015**, *1*, 1–2.

49. Hüttl, R.; Frielinghaus, M. Soil fertility problems—An agriculture and forestry perspective. *Sci. Total Environ.* **1994**,*143*, 63–74.

50. Abbott, L.K.; Murphy, D.V. What is soil biological fertility? In *Soil Biological Fertility—A Key to Sustainable Land Use in Agriculture*; Springer: Berlin/Heidelberg, Germany, 2007.

51. Abiven, S.; Menasseri, S.; Chenu, C. The effects of organic inputs over time on soil aggregate stability. *Soil Biol. Biochem.* **2008**, *41*, 1–12.

52. Li, S.; Wang, Z.; Hu, T.; Gao, Y.; Stewart, B.; Sparks, D. Nitrogen in dryland soils of China and its management. *Adv. Agron.* **2009**, *101*, 123–181.

53. Sherman, K. Sustainability, biomass yields, and health of coastal ecosystems: An ecological perspective. *Marine Ecol. Progr. Series* **1994**, *112*, 277–301.

54. Khaledian, Y.; Kiani, F.; Sohaila, E. The effect of land use change on soil and water quality in northern Iran. *J. Mt. Sci.***2012**, *9*, 798–816.

55. Tsatsaros, J.; Brodie, J.; Bohnet, I.; Valentine, P. Water Quality Degradation of Coastal Waterways in the Wet Tropics, Australia. *Water Air Soil Pollut.* **2013**, *224*.

56. Schaffner, M.; Bader, H.; Scheidegger, R. Modeling the contribution of point sources and non-point sources to Thachin River water pollution. *Sci. Total Environ.* **2009**, *407*, 4902–4915.

57. Atapattu, S.; Kodituwakku, D. Agriculture in South Asia and its implications on downstream health and sustainability: A review. *Agric. Water Manag.* **2009**, *96*, 361–373.

58. Trivedi, R. Water quality of the Ganga River—An overview. *Aquat. Ecosyst. Health Manag.* **2010**, *13*, 347–351.

59. Chen, M.; Chen, J. Phosphorus release from agriculture to surface waters: Past, present and future in China. *Water Sci. Technol.* **2008**, *57*, 1355–1361.

60. Cheevaporn, V.; Menasveta, P. Water pollution and habitat degradation in the Gulf of Thailand. *Marine Pollut. Bull.***2003**, *47*, 43–51.

61. Qadir, M.; Oster, J. Crop and irrigation management strategies for saline-sodic soils and waters aimed at environmentally sustainable agriculture. *Sci. Total Environ.* **2004**, *323*, 1–19.

62. Lemly, A.; Finger, S.; Nelson, M. Sources and impacts of irrigation drainwater contaminants in arid wetlands. *Environ. Toxicol. Chem.* **1993**, *12*, 2265–2279.

63. Dakoure, M.; Mermoud, A.; Yacouba, H.; Boivin, P. Impacts of irrigation with industrial treated wastewater on soil properties. *Geoderma* **2013**, *200*, 31–39.

64. Bouwer, H. Integrated water management for the 21st century: Problems and solutions. *J. Irrig. Drain. Eng.-Asce* **2002**,*128*, 193–202.

65. Apitz, S.E.; Brils, J.; Marcomini, A.; Critto, A.; Agostini, P.; Micheletti, C.; Pippa, R.; Scanferla, P.; Zuin, S.; Lanczos, T.;*et al.* Approaches and frameworks for managing contaminated sediments—A European perspective. In *Assessment and Remediation of Contaminated Sediments*; Springer: Berlin/Heidelberg, Germany, 2006.

66. Lal, R. Societal value of soil carbon. *J. Soil Water Conserv.* **2014**, *69*, 186A–192A.

67. Kohli, R.V.; Singh, H.P.; Batish, D.R.; Jose, S. Ecological interactions in agroforestry: An overview. In *Ecological Basis of Agroforestry*; Kohli, R.V.S., Batish, D.R., Jose, S., Eds.; CRC Press: Boca Raton, FL, USA, 2008; pp. 3–14.

68. Helman, D.; Lensky, I.; Mussery, A.; Leu, S. Rehabilitating degraded drylands by creating woodland islets: Assessing long-term effects on aboveground productivity and soil fertility. *Agric. For. Meteorol.* **2014**, *195*, 52–60.

69. Berazneva, J.; Conrad, J.; Guerena, D.; Lehmann, J. Agricultural productivity and soil carbon dynamics: A bio-economic model. In Proceedings of the Agricultural and Applied Economics Association 2014 Annual Meeting, Minneapolis, MN, USA, 27–29 July 2014.

70. Gong, X.; Liu, Y.; Li, Q.; Wei, X.; Guo, X.; Niu, D.; Zhang, W.; Zhang, J.; Zhang, L. Sub-tropic degraded red soil restoration: Is soil organic carbon build-up limited by nutrients supply. *For. Ecol. Manag.* **2013**, *300*, 77–87.

71. Iqbal, M.; van Es, H.; Anwar-ul-Hassan, R.R.; Schindelbeck, R.; Moebius-Clune, B. Soil Health Indicators as Affected by Long-term Application of Farm Manure and Cropping Patterns under Semi-arid Climates. *Int. J. Agric. Biol.* **2014**,*16*, 242–250

72. Govaerts, B.; Sayre, K.; Deckers, J. A minimum data set for soil quality assessment of wheat and maize cropping in the highlands of Mexico. *Soil Tillage Res.* **2006**, *87*, 163–174.

73. Moebius-Clune, B.; van Es, H.; Idowu, O.; Schindelbeck, R.; Kimetu, J.; Ngoze, S.; Lehmann, J.; Kinyangi, J. Long-term soil quality degradation along a cultivation chronosequence in western Kenya. *Agric. Ecosyst. Environ.* **2011**, *141*, 86–99.

74. Sortino, O.; Montoneri, E.; Patane, C.; Rosato, R.; Tabasso, S.; Ginepro, M. Benefits for agriculture and the environment from urban waste. *Sci. Total Environ.* **2014**, *487*, 443–451.

75. Baumgart-Getz, A.; Prokopy, L.; Floress, K. Why farmers adopt best management practice in the United States: A meta-analysis of the adoption literature. *J. Environ. Manag.* **2012**, *96*, 17–25.

76. Pimentel, D.; Harvey, C.; Resosudormo, P.; Sinclair, K.; Kurz, D.; McNair, M.; Crist, S.; Shpritz, L.; Fitton, L.; Saffouri, R.; *et al.* Environmental and economic costs of soil erosion and conservation benefits. *Science* **1995**, *267*, 1117–1123.

77. Mchunu, C.; Chaplot, V. Land degradation impact on soil carbon losses through water erosion and CO_2 emissions.*Geoderma* **2012**, *177*, 72–79.

78. Badia, D.; Marti, C. Fire and rainfall energy effects on soil erosion and runoff generation in semi-arid forested lands.*Arid Land Res. Manag.* **2008**, *22*, 93–108.

79. Du Preez, C.C.; van Huyssteen, C.W.; Mnkeni, P.N.S. Land use and soil organic matter in South Africa 2: A review on the influence of arable crop production. *South African J. Sci.* **2011**, *107*, 35–42.

80. Longeuville, D.; Henry, S.; Ozer, P. Saharan dust pollution: Implications for the Sahel. *Epidemiologly* **2009**, *20*.

81. Moreno, J.; Bastida, F.; Hernandez, T.; Garcia, C. Relationship between the agricultural management of a semi-arid soil and microbiological quality. *Commun. Soil Sci. Plant Anal.* **2008**, *39*, 421–439.

82. Bastida, F.; Moreno, J.; Hernandez, T.; Garcia, C. Microbiological degradation index of soils in a semiarid climate. *Soil Biol. Biochem.* **2006**, *38*, 3463–3473.

83. Fterich, A.; Mahdhi, M.; Mars, M. Seasonal Changes of Microbiological Properties in Steppe Soils from Degraded Arid Area in Tunisia. *Arid Land Res. Manag.* **2014**, *28*, 49–58.

84. Darwin, C.R. *The Formation of Vegetable Mould, through the Action of Worms, with Observations on their Habitats*; John Murray: London, UK, 1881.

85. Edwards, W.M.; Shipitalo, M.J.; Norton, L.D. Contribution of macropososity to infiltration into a continuous corn no-tilled watershed: Implications for contaminant movement. *J. Contam. Hydrol.* **1988**, *3*, 193–205.

86. Lal, R. *Tropical Ecology and Physical Edaphology*; John Wiley Sons: Chichester, UK, 1987.

87. Castellanos-Navarrete, A.; Rodriguez-Aragones, C.; de Goede, R.; Kooistra, M.; Sayre, K.; Brussaard, L.; Pulleman, M. Earthworm activity and soil structural changes under conservation agriculture in central Mexico. *Soil Tillage Res.* **2012**,*123*, 61–70.

88. Ayuke, F.; Karanja, N.; Okello, J.; Wachira, P.; Mutua, G.; Lelei, D.; Gachene, C.; Hester, R.; Harrison, R. Agrobiodiversity and Potential Use for Enhancing Soil Health in Tropical Soils of Africa. *Soils Food Secur.* **2012**, *35*, 94–134.

89. Ryan, J.; Masri, S.; Ibrikci, H.; Singh, M.; Pala, M.; Harris, H. Implications of cereal-based crop rotations, nitrogen fertilization, and stubble grazing on soil organic matter in a Mediterranean-type environment. *Turkish J. Agric. For.***2008**, *32*, 289–297.

90. Teague, W.; Foy, J.; Cross, B.; Dowhower, S. Soil carbon and nitrogen changes following root-plowing of rangeland. *J. Range Manag.* **1999**, *52*, 666–670.

91. Emmerich, W.; Heitschmidt, R. Drought and grazing: II. Effects on runoff and water quality. *J. Range Manag.* **2002**, *55*, 229–234.

92. Snyman, H.; du Preez, C. Rangeland degradation in a semi-arid South Africa—II: Influence on soil quality. *J. Arid Environ.* **2005**, *60*, 483–507.

93. Oesterheld, M.; Loreti, J.; Semmartin, M.; Sala, O. Inter-annual variation in primary production of a semi-arid grassland related to previous-year production. *J. Veg. Sci.* **2001**, *12*, 137–142.

94. Wiegand, T.; Snyman, H.; Kellner, K.; Paruelo, J. Do grasslands have a memory: Modeling phytomass production of a semiarid South African grassland. *Ecosystems* **2004**, *7*, 243–258.

95. Thapa, G.; Yila, O. Farmers land management practices and status of agricultural land in the Jos Plateau, Nigeria. *Land Degrad. Dev.* **2012**, *23*, 263–277

96. Garrity, D.; Akinnifesi, F.; Ajayi, O.; Weldesemayat, S.; Mowo, J.; Kalinganire, A.; Larwanou, M.; Bayala, J. Evergreen Agriculture: A robust approach to sustainable food security in Africa. *Food Secur.* **2010**, *2*, 197–214.

97. Muir, J.; Pitman, W.; Foster, J. Sustainable, low-input, warm-season, grass-legume grassland mixtures: Mission (nearly) impossible? *Grass Forage Sci.* **2011**, *66*, 301–315.

98. Lal, R. Degradation and resilience of soils. *Phil. Trans. R. Soc. Lond. B.* **1997**, *352*, 997–1010.

99. Greenland, D.J., Szabolcs, I., Eds.; *Soil Resilience and Sustainable Land Use*; CAB International: Wallingford, UK, 1994.

100. Lynch, J. Resilience of the rhizosphere to anthropogenic disturbance. *Biodegradation* **2002**, *13*, 21–27.

101. Syers, J.K. Manging soils for long-term productivity. *Philos. Trans. R. Lond. B* **1997**, *352*, 1011–1021.

102. Horrigan, L.; Lawrence, R.; Walker, P. How sustainable agriculture can address the environmental and human health harms of industrial agriculture. *Environ. Health Perspect.* **2002**, *110*, 445–456.

103. Breulmann, M.; van Afferden, M.; Führer, C. Biochar: Bring on the sewage. *Nature* **2015**, *518*, 483.

104. Goodman-Elgar, M. Evaluating soil resilience in long-term cultivation: A study of pre-Columbian terraces from the Paca Valley, Peru. *J. Archaeol. Sci.* **2008**, *35*, 3072–3086.

105. Ogaji, J. Sustainable agriculture in the UK. *Environ. Dev. Sustain.* **2005**, *7*, 253–270.

106. Ravnborg, H. Poverty and soil management—Relationships from three Honduran watersheds. *Soc. Nat. Resour.* **2002**,*15*, 523–539.

107. Tittonell, P.; Giller, K. When yield gaps are poverty traps: The paradigm of ecological intensification in African smallholder agriculture. *Field Crops Res.* **2013**, *143*, 76–90.

108. Rozelle, S.; Huang, J.; Zhang, L. Poverty, population and environmental degradation in China. *Food Policy* **1997**, *22*, 229–251.

109. Fan, M.; Shen, J.; Yuan, L.; Jiang, R.; Chen, X.; Davies, W.; Zhang, F. Improving crop productivity and resource use efficiency to ensure food security and environmental quality in China. *J. Exp. Bot.* **2012**, *63*, 13–24.

110. Tennesen, M. Rare Earth. *Science* **2014**, *346*, 692–695.

111. Reuters. Peak soil threatens global food security. Reuters, 17 July 2014.

112. Flora, C. Food security in the context of energy and resource depletion: Sustainable agriculture in developing countries. *Renew. Agric. Food Syst.* **2010**, *25*, 118–128.

113. Wall, D.; Six, J. Give soils their due. *Science* **2015**, *347*, 695.

114. Gregory, P.; Hester, R.; Harrison, R. Soils and Food Security: Challenges and Opportunities. *Soils Food Secur.* **2012**, *35*, 1–30.

Chapter 2

Soil Fertility, Phosphorus Fractions, and Maize Yield as Affected by Poultry Manure and Single Superphosphate

A. O. Ojo,[1] M. T. Adetunji,[2] K. A. Okeleye,[2] and C. O. Adejuyigbe[2]

[1]Institute of Agricultural Research and Training, Obafemi Awolowo University, P.M.B 5029, Moor Plantation, Ibadan, Oyo State, Nigeria

[2]Federal University of Agriculture, Abeokuta, Nigeria

ABSTRACT

A field experiment was conducted in 2007 and 2008 on a slightly acidic alfisol. Poultry manure (PM) was applied at 0, 5 t ha^{-1}, 10 t ha^{-1}, 15 t ha^{-1}, and 20 t ha^{-1} in combination with SSP at 0, 15 kg P ha^{-1}, 30 kg P ha^{-1}, 45 kg P ha^{-1}, and 60 kg P ha^{-1}, which was replicated three times. The pH and organic C were significantly increased by the application of PM alone while available P was highly increased by the sole application of SSP. Plant tissue P was significantly increased with the application of 30 kg P ha^{-1} while the largest grain yield was obtained when PM at 20 t ha^{-1} was combined with SSP at 60 kg P ha^{-1}. The buildup of organic P was observed when PM was applied at 15 t ha^{-1} while the combination of the two treatments increased residual P and Fe-P. However, P occlusion was effectively reduced with the sole application of PM. Organic P and residual P however had a strong positive relationship with the grain yield. Comparing the sole and combined application of the treatments, the combined application was more effective for most of the parameters observed.

1. INTRODUCTION

Phosphorus (P) is one of the essential elements for plant growth and the number of plants available P in the soil is often inadequate to meet plant requirements. Phosphorus is a macronutrient that plays a number of important roles in plant. Adequate phosphorus results in higher grain production, improved crop quality, greater stalk strength, increased root growth, and earlier crop maturity. Mokwunye et al. [1] and Warren [2] stated that phosphorus deficiency is one of the largest constraints to food production in tropical African soils due to low native P and high fixation by iron and aluminum oxides. It has also been established that phosphorus is relatively unavailable for plant uptake in highly weathered soils like the Ultisols [3]. In order to meet the need of phosphorus in these soils, farmers make use of inorganic fertilizers for their crops. However, P fertilizer is not readily available to these farmers because of scarcity in some cases and also due to the high cost of these fertilizers. Recently, farmers especially in the developed countries make use of organic manure which has been found to be effective like inorganic fertilizers for the release of some nutrients especially phosphorus but the resultant yield has been found not to be as high as when inorganic fertilizer is combined with the organic manure. Combining organic manure with P fertilizers has been found to not only increase yield but also reduce P fixation and subsequently increase phosphorus availability and uptake [3]. The study however aims at studying the effect of combined application of poultry manure and single superphosphate on phosphorus forms, soil fertility, and the yield maize.

2. MATERIALS AND METHODS

The field experiment was conducted at I.A.R&T experimental field at Ibadan in 2007 and 2008 early planting season (see Table 1). The experimental design was a 5 × 5 factorial experiment in a RCBD replicated three times. The plot size was 5 m × 4 m (20 m^2) with spacing of 75 cm × 50 cm to give a total plant population of 106,666 plants ha^{-1}. Three seeds of maize (TZB-SR) were planted after 2 weeks of treatment incorporation. The seedlings were thinned to two per stand two weeks after planting. The plots were not tilled in the early season (May) for the second year (2008) in order to avoid mixing up of treatments. Gramoxone and primextra were used as herbicides. Weeding was done at the 4th and 8th week after planting.

Table 1. Source of the soil used.

Location	State	Coordinates	Vegetation	Soil series	Soil class (USDA)
Ibadan	Oyo	Lat. 7°30′ Long. 3°54′	Rainforest	Iwo	Ferrudulf

Leaf sampling of maize was done at eight weeks after planting. Leaf just below and opposite the ear was sampled randomly per plot. Ears of maize were allowed to dry on the stalk. Ears in the four middle rows of each plot were harvested (making allowance for two guard rows) on each side of the plot. The cobs were dehusked and weighed. Ten cobs per plots were randomly selected for cob weight measurement. The cobs were sun dried and the grains were weighed. Yield per hectare was computed at 12% moisture content for maize. Soil samples were analyzed before and after for phosphorus fractions and soil nutrient composition.

2.1. Soil Physical and Chemical Analysis

Particle size distribution was determined by the hydrometer method [4] using sodium hexametaphosphate as the dispersing agent. Soil pH was determined in distilled water (1: 1 soil water ratio) with a pH meter. Exchangeable cations (K, Na, Ca, and Mg) by extraction, with ammonium acetate (pH = 7). Available P was extracted with 0.03 N NH_4F in 0.025 N HCl solution [5] and P in the extract was analysed colorimetrically by the molybdenum blue method at 660 nm. Soil organic carbon was determined by wet oxidation with sulphuric acid [6]. Total P was determined by wet digestion using perchloric acid and nitric acid while organic P was by ignition method. Phosphorus concentration in the digests was determined using the ascorbic acid method.

Table 2. Sequential P fractionation procedures and targeted P forms.

Location	Extractants	Equilibration	Washing	Targeted
NH_4Cl-P	IM NH_4Cl	30 mins	None	Saloid-bound P (loosely bound P)
NH_4F-P	0.5 M NH_4F	1 hour	None	Al-P
NaOH-P	0.1 M NaOH saturated Nacl + 5 drops of conc. H_2SO_4	15 mins	Saturated Nacl	Fe-P
Sodium + citrate Sodium dithionite P	0.3 M sodium citrate + 1 g solid Sodium dithionate + sat Nacl	Shake 15 mins, preheat 15 mins at 85°C after sodium citrate, additional 15 mins after dithionite addition	Saturated Nacl	Reductant-soluble P
NaOH-P	0.1 M NaOH + few drops of conc. H_2SO_4 to remove colour	15 mins	Saturated NaCl	Occluded P
H_2SO_4-P	0.25 M H_2SO_4 + 25 mL Sat. Nacl	1 hour	Saturated Nacl	Ca-P
Residual P	Conc. HNO_3 + HCl + 30% H_2O_2	Variable until complete	None	Organically stable organic and inorganic P

2.2. Phosphorus Fractionation Study

Fractionation procedure by Chang and Jackson [7] was used to quantify different forms of inorganic P in the studied soils (Table 2). Fractionation was done on the soil before and after amendment for incubation, greenhouse, and field experiment. The sequence of the chemical extraction and a brief description of the targeted P forms are shown in Table 2.

2.3. Determination of Phosphorus in the Extracts

The phosphorus in the extracts was determined by the methyl blue colour method of Murphy and Riley [8]. An aliquot of 10 mL was used in developing the colour. The determination of P in NH_4F, NaOH, and DCB extracts however requires the following pretreatment before the development of the colour. (a)NH_4F extract: 0.05 g of H_3BO_3 per 50 mL of the extract was added before taking the aliquot. (b)NaOH extract: after precipitating the organic matter with conc. H_2SO_4, the pH of the extract was adjusted with NaOH using 2,4-dinitrophenol indicator. An aliquot of the clear solution was then taken into a 50 mL volumetric flask.(c)DCB extract (reductant soluble P): 10 mL of the extract was pipetted into a test tube and 1 mL of $HClO_4$was added to it and boiled to destroy the dithionite in the extract. It was then transferred into a 50 mL volumetric flask and 3 mL of 5% ammonium molybdate solution was added while distilled water was used to make up to mark. Absorbance was read after 30 minutes.

2.4. Analysis of Poultry Manure Sample

Total analysis of the sample was done by weighing 0.5 g of poultry manure, adding 10 mL of $HNO_3/HClO_4$, (2 : 1), and digesting at 150°C. The temperature was increased at 1.5 hours to 230°C and 2 mL of HCl/H_2O (1 : 1) was added. The digestion was now continued for another 30 minutes [9].

2.5. Statistical Analysis

The data collected were subjected to analysis of variance using the statistical analysis system (SAS)—general linear model [10]. Means were separated by Duncan multiple range test.

3. RESULTS

3.1. Initial Chemical Properties of the Soil Used

The pH of the soil was slightly acidic and therefore a need for the application of organic manure. The soil was low in N, Na, and organic carbon while P was moderately available in it. It was however adequate with cations like Ca, Mg, and K (Table 3).

Table 3. Chemical and physical properties of the soil used for the study.

Parameters	
pH (H_2O)	6.4
O.M ($g\,kg^{-1}$)	24.08
N ($g\,kg^{-1}$)	0.2
P ($mg\,kg^{-1}$)	14.2
Exchangeable cations ($c\,molc\,kg^{-1}$)	
Ca	1.14
Mg	0.82
K	0.47
Na	0.09
Particle size ($g\,kg^{-1}$)	
Sand	740
Clay	120
Silt	140

3.2. Characterization of the Poultry Manure Used and the Initial Phosphorus Fractions of the Soil

Nitrogen in the poultry manure used was 5.82% which was similar to the result observed for phosphorus (5.54%) while potassium was low in the soil (0.94%). Organic C had a high percentage of 13.38% which led to the resultant C/N ratio of 2.3 (Table 4). Also notable was the result obtained for Fe (1555 mg/kg).

Table 4. Characterization of the poultry manure used.

pH	7.03
Nitrogen (N) (%)	5.82
Phosphorus (P) (%)	5.54
Potassium (K) (%)	0.94
Na (%)	1.3
SO_4-S (%)	0.14
Organic C (%)	13.34
C/N	2.3
Calcium (%)	8.04
Magnesium (%)	0.61
Iron (mg kg^{-1})	1555
Copper (mg kg^{-1})	33.3
Zinc (mg kg^{-1})	100.3
Manganese (mg kg^{-1})	180

The order of abundance of the phosphorus fractions before the application of treatments was organic P Fe-P ≡ occluded P residual P (Table 5).

Table 5. Phosphorus fractions of the soil before planting (mg kg^{-1}).

Phosphorus fractions (mg kg^{-1})	
Total P	272.84
Organic P	45.90
Fe-P	33.39
Occluded P	33.39
Residual P	16.16

3.3. Effect of Combined Application of Poultry Manure and SSP on the Soil Chemical Properties

There was a decrease in pH as a result of the sole application of SSP in 2007, which further decreased in 2008 (Table 6). Application of poultry manure alone increased the pH of the soil significantly when compared to the other treatments. However, the combined application of poultry manure and SSP increased the pH of the soil in 2007 with a further increase in 2008. In other words, poultry manure alone can be compared favorably with the combined application of poultry manure and SSP in increasing soil pH especially in slightly acidic soil. Effect of SSP on organic C was not significant and decreased when compared to the initial value. However, the application of poultry manure increased organic C significantly when poultry manure was applied alone in 2007. In 2008, organic C further increased with the sole application of poultry manure. The effect of the combined application of poultry manure and SSP was not significant on organic carbon in 2007 but significant in 2008 with the combined application of poultry manure and SSP. Available P increased over the control either with the sole application of poultry manure or SSP or when the two treatments were combined. However, in 2007 the sole application of SSP was more effective in increasing available P than the sole application of poultry manure but was the most effective with the combined application of poultry manure and SSP, while, in 2008, the combined treatments were not significant but the sole application was.

Table 6. Effect of poultry manure (PM) and SSP on some soil chemical properties in 2007 and 2008.

Treatments PM (t ha^{-1})	SSP (kg P ha^{-1})	pH (2007)	pH (2008)	Organic C, % (2007)	Organic C, % (2008)	Available P, mg kg^{-1} (2007)	Available P, mg kg^{-1} (2008)
	0	6.23	6.30	0.19	0.16	5.90	6.07
	15	6.73	6.73	0.62	0.85	92.57	96.78
0	30	6.32	6.30	1.02	1.27	39.41	39.91
	45	6.51	6.51	1.26	0.74	69.60	70.02
	60	6.42	6.87	0.75	0.89	42.56	41.18
	0	7.06	7.06	0.65	1.05	41.50	41.70
	15	6.53	6.53	0.92	1.90	56.39	56.27
5	30	6.52	6.52	1.06	1.42	43.56	43.84
	45	6.65	6.62	1.12	1.29	90.05	87.66
	60	6.92	6.92	0.91	0.97	64.69	65.82
	0	6.87	6.87	1.03	1.39	64.82	63.30
	15	6.82	6.73	1.03	1.04	59.62	59.74
10	30	6.72	6.72	1.37	0.87	67.06	66.74
	45	6.86	6.86	0.98	1.34	83.32	81.65
	60	6.87	6.87	1.02	1.53	60.42	60.42
	0	6.50	6.50	1.63	1.52	40.05	96.78
	15	6.87	6.93	1.66	2.65	67.71	67.78
15	30	6.80	6.80	1.08	1.45	96.82	87.49
	45	6.65	6.72	1.08	1.62	88.31	87.24
	60	6.62	7.16	1.25	1.50	75.57	74.54

	0	6.93	6.93	1.03	1.75	59.60	53.07
	15	6.60	6.60	0.78	1.01	72.68	76.81
20	30	6.60	6.60	1.22	1.26	72.68	79.69
	45	6.50	6.50	0.82	1.12	64.87	64.44
	60	6.97	6.96	1.23	2.60	61.74	94.84
	LSD ($P \leq 0.05$)						
	SSP	*	*	NS	*	*	NS
	PM	*	*	*	*	*	NS
	SSP × PM	*	NS	NS	*	*	NS

*Significant at $P < 0.05$, NS: not significant.

3.4. Effect of the Combined Application of Poultry Manure and SSP on Plant Tissue P Concentration and Grain Yield of Maize

Effect of poultry manure and SSP alone with their combinations was significant in both years (Table 7). In 2007, plant tissue P was high when the treatments were applied solely but combined application was highly significant. Plant tissue P continued to increase in 2008 with the sole and combined application but with the most significant increase with the sole application of SSP.

The sole application of SSP had no significant effect on the grain yield of maize in 2007 while the sole application of poultry manure and the combined application of the two treatments were significant in the same year. However, in 2007, the highest grain yield of maize was obtained by the sole application of poultry manure. In 2008, effect of the sole application of poultry manure and SSP on maize grain yield was significant while the combined application was not. The combined application of the two treatments gave the highest grain yield in 2008.

3.5. Effect of the Combined Application of Poultry Manure and SSP on Phosphorus Fractions

Effect of the sole and combined application of poultry manure and SSP was significant in both 2007 and 2008 (Table 8). In 2007, organic P was significantly increased over the initial value when poultry manure was applied solely than when SSP was applied alone. Combining the two treatments also increased organic P significantly in 2007. The application of poultry manure alone however gave the highest increase in organic P in 2007. In 2008, organic P reduced with the sole application of SSP but increased when poultry manure was applied alone and with some combined treatments. The sole application of poultry manure was again effective in 2008 in increasing organic P. Residual P increased over the control when poultry manure and SSP were applied alone and when combined in both years. Residual P increased in 2007 with the combined application of the two treatments but reduced in 2008. A similar trend to that of residual P was observed for Fe-P but with few exceptions when the 2 treatments were combined. In 2008, the largest buildup of Fe-P was observed with the combined application of poultry manure and SSP. In both years of cropping, sole application of poultry manure was effective in reducing P occlusion more than the combined application.

3.6. Relationship among the P Fractions, Soil Chemical Properties, and Maize Grain Yield

In 2007, the pH of the soil had a strong negative relationship with Al-P and occluded P but positive relationship with residual P (Table 9), while in 2008, increase in the pH of the soil favored the buildup of Fe-P (Table 10). Available P in 2007 had a positive correlation with organic P, Ca-P, and Fe-P but a negative correlation with occluded P. Organic C had no significant relationship with the P fractions in 2007 but a positive correlation with organic P and a negative correlation with occluded P in 2008. Organic P and residual P contributed positively to the maize grain yield in both 2007 and 2008.

4. DISCUSSION

4.1. Initial Chemical Properties of the Soil Used

The pH of the soil was slightly acidic and this will make Fe-P be more available than the other inorganic P fractions. Although phosphorus was moderately available, poultry manure and SSP were further applied in order to study the effect of continuous application of organic and inorganic fertilizer on such soils yearly as has been practiced by many farmers in Nigeria.

4.2. Characterization of the Poultry Manure Used and Initial Phosphorus Fractions of the Soil

A common and traditional practice has utilized manure as an organic fertilizer or soil improver due to its contribution of organic matter and nutrients, mainly N and P [11]. The high percentage of nitrogen in the poultry manure used could have been due to the decomposition of poultry manure during the curing stage before application to the soil with a subsequent release of N which was detected during the analysis. Manure in past studies has been applied according to the nitrogen needs of the soil [12]. Macronutrients such as phosphorus are part of the advantage of using organic waste in agriculture, which may be recovered and incorporated into natural cycle and this was evident in the percentage of phosphorus in the poultry manure used. It also contains a high level of organic matter which generates changes in chemical, physical, and biological soil amendment properties [12].

Organic P was the highest P fraction at the initial stage of this study. About two-thirds of the phosphorus in fresh manure is in organic form [12]. The organic P pool is usually considered a major source of organic phosphorus and can be converted to inorganic P by ignition or wet digestion. Initial abundance of Fe-P could be attributed to the acidic nature of the soil; however occluded P also had the same value which could mean there is an equilibrium reaction between the labile and the nonlabile P fractions.

4.3. Effect of Combined Application of Poultry Manure and SSP on the Soil Chemical Properties

Acidity in sols if not properly managed will eventually make the soil be low or not yielding and therefore a need for amendment. In this study, application of poultry manure compared favourably with the combined application of poultry manure and SSP in increasing the pH of the soil while applying inorganic fertilizer alone decreased the soil pH. Soil pH of an acidic Ultisol in the eastern Nigeria was observed to increase soil pH [13]. Sole and combined organic and inorganic fertilizer has been found to increase soil nutrients favorably. The combined application of the two treatments was highly effective in increasing soil organic C in both 2007 and 2008. Earlier studies on poultry manure ability to increase soil organic C have been reported [13, 14], likewise when it is combined with SSP [15]. Single superphosphate was however more effective than the sole application of poultry manure for increasing the available P content of the soil although the combined application of the two treatments still gave the highest value [16]. The effectiveness of SSP for available P over poultry manure could be due to the immediate release of phosphate into the soil when SSP is being applied.

Table 7. Effect of poultry manure (PM) and SSP on plant tissue concentration and grain yield in 2007 and 2008.

Treatments		Plant tissue (%) P		Grain yield (t ha^{-1})	
PM (t ha^{-1})	SSP (kg P ha^{-1})	2007	2008	2007	2008
0	0	0.43	0.53	1.33	0.81
	15	0.94	1.04	3.12	3.21
	30	0.82	2.51	2.29	1.45
	45	1.29	2.02	2.57	1.05
	60	0.79	1.93	2.39	1.42
5	0	0.88	0.77	2.76	1.98
	15	1.97	1.56	2.61	1.25
	30	0.87	0.96	2.69	1.67
	45	0.85	2.50	2.71	1.74
	60	0.82	1.24	2.35	2.57
10	0	1.32	0.86	2.87	1.66
	15	1.49	1.82	2.46	2.18
	30	1.01	1.33	2.59	3.02
	45	1.61	0.84	2.48	1.78
	60	1.28	1.35	2.42	3.53
15	0	2.00	1.04	3.63	0.83
	15	1.82	1.01	2.75	3.85
	30	0.87	0.83	2.75	3.50
	45	1.49	1.42	2.60	3.21
	60	0.93	1.02	2.20	3.76

	0	1.03	0.94	2.82	2.82
	15	0.87	1.36	2.86	2.41
20	30	1.36	1.52	2.64	2.44
	45	0.99	2.09	2.36	2.30
	60	0.99	1.35	2.37	3.99
	LSD ($P \leq 0.05$)				
	SSP	*	*	NS	*
	PM	*	*	*	*
	SSP × PM	*	*	*	NS

* Significant at $P < 0.05$, NS: not significant.

Table 8. Effects of poultry manure and SSP on phosphorus fractions in 2007 and 2008.

Treatments		Phosphorus fractions (mg kg^{-1})							
PM (t ha^{-1})	P rates (Kg P ha^{-1})	Organic P 2007	Organic P 2008	Residual P 2007	Residual P 2008	Fe-P 2007	Fe-P 2008	Occluded P 2007	Occluded P 2008
0	0	49.91	50.04	21.83	12.26	38.50	19.48	27.25	28.03
	15	128.96	112.50	36.41	18.55	88.21	21.62	31.61	10.11
	30	135.63	125.02	36.08	17.54	90.24	17.74	22.66	10.53
	45	115.63	140.63	40.73	20.52	68.09	64.92	31.65	11.72
	60	177.50	130.58	54.99	22.97	117.37	18.86	36.72	12.24
5	0	75.03	111.27	47.43	17.45	88.12	50.76	16.61	9.36
	15	138.75	168.75	37.18	15.52	86.82	16.72	19.79	11.08
	30	167.35	146.85	41.47	17.34	88.44	16.85	26.08	9.56
	45	143.14	116.22	48.76	19.31	87.08	137.76	27.82	12.34
	60	100.06	106.83	46.26	19.22	116.29	62.25	21.18	17.14
10	0	105.07	160.05	49.32	17.43	87.15	86.81	18.63	10.52
	15	180.05	118.75	45.24	16.45	88.31	19.07	26.82	8.51
	30	183.73	155.05	51.29	18.63	115.23	15.35	26.18	10.33
	45	160.37	122.50	38.16	21.28	104.62	117.55	44.02	12.88
	60	145.61	130.11	42.28	20.75	117.69	148.94	30.12	11.72
15	0	198.75	112.50	36.61	16.92	61.81	21.62	12.14	17.01
	15	91.85	120.02	45.57	20.07	97.69	18.25	22.89	8.51
	30	124.38	159.38	41.89	18.62	117.68	15.96	38.42	10.64
	45	105.53	174.33	40.11	25.54	98.54	131.93	42.11	14.26
	60	161.86	170.52	34.45	22.76	202.36	152.11	28.95	9.92
20	0	149.35	103.77	47.51	18.91	84.52	16.08	12.46	12.13
	15	99.39	101.88	44.41	16.07	75.13	18.95	32.88	9.37
	30	141.85	174.83	42.39	21.05	117.99	15.97	34.91	11.08
	45	195.00	136.22	54.63	31.62	89.62	13.18	36.81	11.21
	60	171.25	150.02	57.24	23.65	113.12	125.41	39.37	13.21
	LSD ($P \leq 0.05$)								
	SSP	*	*	*	*	*	*	*	*
	PM	*	*	*	*	*	*	*	*
	SSP × PM	*	*	*	*	*	*	*	*

*Significant at $P < 0.05$, NS: not significant.

Table 9. Correlation coefficient (*r*) among P fractions, soil chemical properties, and maize grain yield in 2007.

Soil properties	P fractions (mg kg^{-1})			
	Organic P	Fe-P	Occluded P	Residual P
pH	0.098	0.201	−0.550	0.633
Available P	0.514	0.433	−0.528	−0.071
Organic C	0.206	0.115	−0.031	−0.105
Grain yield	0.413	0.145	0.034	0.402

Table 10. Correlation coefficient (*r*) among P fractions, soil chemical properties, and maize grain yield in 2008.

Soil properties	P fractions (mg kg^{-1})			
	Organic P	Fe-P	Occluded P	Residual P
pH	0.118	0.458	0.065	−0.144
Available P	0.053	0.173	−0.753	0.048
Organic C	0.500	0.122	−0.557	−0.177
Grain yield	0.403	0.105	0.123	0.412

4.4. Effect of the Combined Application of Poultry Manure and SSP on Plant Tissue P Concentration and Grain Yield of Maize

High plant tissue P was obtained in 2007 and 2008 with the sole and combined application of poultry manure and SSP. However, the sole application of SSP was however sustainable in the 2nd year to cause a further increase in the concentration of P in the shoot. A similar result has been observed by Adeniyan and Ojeniyi, [17]. The increase in the plant tissue P in 2008 was as a result of P accumulation observed in both years, which would have resulted in the transfer to the shoot of the plant instead of being washed into nearby river and hereby preventing eutrophication.

When the treatments were applied solely, application of poultry manure alone was more effective than SSP, that is, considering the high grain yield. Although this was in the first year of planting, the combined application of poultry manure and SSP was again sustainable in the second year. The combined application of organic and inorganic resources has been found to favor maize production [14].

4.5. Effect of the Combined Application of Poultry Manure and SSP on Phosphorus Fractions

The buildup of organic P was observed in both years of this study with the sole application of poultry manure and this signifies the importance of organic manure when considering amendments for an acidic soil. Addition of soybean residue and wheat residue or in combination with fertilizer P favored the buildup of organic P in a study [18]. The inorganic P fractions considered, that is, residual P and Fe-P, was significantly increased when poultry manure was combined with SSP. This indicates the ability of the combined application of poultry manure and SSP to maintain the fertility of the soil through increase in the labile P fraction [19]. Residual P however reduced in 2008 and this was also observed by Agbenin and Goladi, [20] in an experiment involving thc addition of P fertilizers or in combination with cow manure. Significant was the ability of poultry manure alone to reduce P occlusion in both years, that is, occluded P, which further confirms an earlier study which states that addition of organic materials to soil decreases P fixation [3]. Inorganic and organic products are generated during the partial decomposition of organic waste and organic acids can be adsorbed into the soil surface thereby decreasing the potential P adsorption by blocking the formation of complexes [21].

4.6. Relationship among the P Fractions, Soil Chemical Properties, and Maize Grain Yield

Organic P and Fe-P contributed significantly to the increase observed in available P in 2007 while occluded P decreased with an increase in available P. This further confirms the importance of poultry manure while considering soil amelioration. In both years of cropping, organic P was sustainable in increasing grain yield but Fe-P was only effective in 2007. Residual P (a nonlabile P fraction) equally contributed to the increase observed in grain yield in both years. In other words, there would have been an equilibrium reaction between the nonlabile P (residual P) and labile P (Fe-P) fraction which would have subsequently resulted in an increase in available P.

5. CONCLUSION

Application of poultry manure alone was effective in increasing the pH, organic C, and available P content of the soil. However, application of SSP was highly significant in increasing the available P content of the soil and this was evident in the increase obtained for the plant tissue P in 2008, while, for grain yield, the most effective was the combined application of poultry manure and SSP. Increase in organic P fraction was as a result of the application of poultry manure while the combined application of poultry manure and SSP increased the Fe-P fraction. Poultry manure alone was also the most effective treatment for the reduction of P occlusion in the soil. Significant were also the importance of poultry manure and the combined application of poultry manure and SSP exhibited through the strong relationship that existed among available P, Fe-P, and organic P. Although poultry manure was compared favorably with the combined application of poultry manure and SSP, the combination of the treatments was more suitable for the sustainability of the fertility of the slightly acidic soil.

REFERENCES

1. A. U. Mokwunye, S. H. Chien, and E. Rhodes, “Phosphorus reaction with tropical African soils,” inManagement of Nitrogen and Phosphorus Fertilizers in Sub-Saharan African, A. U. Mokwanye and P. L. C. Vick, Eds., pp. 253–281, Martins Nulhoff, Dordrecht, The Netherlands, 1986.

2. G. Warren, Fertilizer Phosphorus Sorption and Residual Value in Tropical African Soils, NRI Bulletin 37, Natural Resource Institute, Chatham, UK, 1992.

3. F. Iyamuremye, R. P. Dick, and J. Baham, "Organic amendments and phosphorus dynamics: I. Phosphorus chemistry and sorption," Soil Science, vol. 161, no. 7, pp. 426–435, 1996.

4. G. J. Bouyoucos, "Hydrometer method improved for making particle size analysis of soils," Agronomy Journal, vol. 54, no. 5, pp. 464–465, 1962.

5. R. H. Bray and L. T. Kurtz, "Determination of total, organic and available forms of phosphorus in soils,"Soil Science, vol. 59, pp. 39–45, 1945.

6. A. Walkey and I. A. Black, "An examination of the Degtjareff method for determining organic carbon in soils: effect of variations in digestion conditions and of inorganic soil constituents," Soil Science, vol. 63, pp. 251–263, 1934.

7. S. C. Chang and M. L. Jackson, "Fraction of soil phosphates," Soil Science, vol. 83, pp. 133–144, 1957.

8. J. Murphy and J. P. Riley, "A modified single solution method for the determination of phosphate in natural waters," Analytica Chimica Acta, vol. 27, pp. 31–36, 1962.

9. E. J. Udo and J. A. Ogunwale, Manual of Soil, Plant and Water Analysis, Department of Agronomy, University of Ibadan, Ibadan, Nigeria, 1978.

10. SAS Institute, SAS for Linear Models. A Guide to the ANOVA and GLM Procedures, SAS Institute, Cary, NC, USA, 2000.

11. Z. He, T. S. Griffin, and C. W. Honeycutt, "Evaluation of soil phosphorus transformations by sequential fractionation and phosphatase hydrolysis," Soil Science, vol. 169, no. 7, pp. 515–527, 2004.

12. B. J. Wienhold, "Changes in soil attributes following low phosphorus swine slurry application to no-tillage sorghum," Soil Science Society of America Journal, vol. 69, no. 1, pp. 206–214, 2005.

13. A. O. Ano and J. A. Agwu, "Effects of animal manures on selected soil chemical properties," Nigeria Journal of Soil Science, vol. 15, pp. 14–19, 2005.

14. L. R. Bulluck III, M. Brosius, G. K. Evanylo, and J. B. Ristaino, "Organic and synthetic fertility amendments influence soil microbial, physical and chemical properties on organic and conventional farms," Applied Soil Ecology, vol. 19, no. 2, pp. 147–160, 2002.

15. M. C. Manna, A. Swarup, R. H. Wanjari, B. Mishra, and D. K. Shahi, "Long-term fertilization, manure and liming effects on soil organic matter and crop yields," Soil and Tillage Research, vol. 94, no. 2, pp. 397–409, 2007.

16. H. Zhang, B. Wang, and M. Xu, "Effects of inorganic fertilizer inputs on grain yields and soil properties in a long-term wheat-corn cropping system in South China," Communications in Soil Science and Plant Analysis, vol. 39, no. 11-12, pp. 1583–1599, 2008.

17. O. N. Adeniyan and S. O. Ojeniyi, "Effect of poultry manure, NPK 15-15-15 and combination of their reduced levels on growth and soil chemical properties," Nigeria Journal of Soil Science, vol. 15, pp. 34–41, 2005.

18. D. D. Reddy, S. A. Rao, and M. Singh, "Changes in P fractions and sorption in an Alfisol following crop residues application," Journal of Plant Nutrition and Soil Science, vol. 168, no. 2, pp. 241–247, 2005.

19. T. Q. Zhang and A. F. MacKenzie, "Changes of soil phosphorous fractions under long-term corn monoculture," Soil Science Society of America Journal, vol. 61, no. 2, pp. 485–493, 1997.

20. J. O. Agbenin and J. T. Goladi, "Dynamics of phosphorus fractions in a savanna Alfisol under continuous cultivation," Soil Use and Management, vol. 14, no. 2, pp. 59–64, 1998.

21. M. S. Mkhabela and P. R. Warman, "The influence of municipal solid waste compost on yield, soil phosphorus availability and uptake by two vegetable crops grown in a Pugwash sandy loam soil in Nova Scotia," Agriculture, Ecosystems and Environment, vol. 106, no. 1, pp. 57–67, 2005.

Chapter 3

Integrated Soil Fertility Management: Impact of Mucuna and Tithonia Biomass on Tomato (Lycopersicon esculentum L.) Performance in Smallholder Farming Systems

Christopher Ngosong[1*], Priscilla M. Mfombep[2], Akeme C. Njume[1], Aaron S. Tening[1]

[1]Department of Agronomic and Applied Molecular Sciences, Faculty of Agriculture and Veterinary Medicine, University of Buea, Buea, Cameroon

[2]Department of Agriculture, Higher Technical Teachers' Training College Kumba, University of Buea, Buea, Cameroon

ABSTRACT

Many views, paradigms and concepts have been advocated in recent decades on soil fertility and soil conservation across the globe in order to provide sustainable solutions to the rising food and nutrition insecurity while preserving the natural resource base. Meanwhile, food and nutrition security in Sub-Saharan Africa (SSA) is mainly achieved through smallholder farming systems that are characterized by poor and declining soil fertility, which often leads to low crop yields and low income. Hence, a field trial was established to evaluate the impact of integrated soil fertility management (ISFM) practices on tomato yield and the farm-scale income in smallholder farming systems. The ISFM trial comprised a control with no input, mineral fertilizer, and organic treatments comprising sole Mucuna and Tithonia biomasses as well as their combination (Mucuna + Tithonia). Generally, tomato performance was better with organic plant biomass amendments, with significantly higher ($P < 0.001$) tomato yield for Mucuna + Tithonia and sole Tithonia, followed by sole Mucuna and mineral fertilizer compared to the control. Meanwhile in comparison to the control, Mucuna + Tithonia and sole Tithonia recorded 3.5 and 3.4 t ha^{-1} more yield, respectively, which was about twice the additional yield for sole Mucuna and mineral fertilizer with 1.8 and 1.5 t ha^{-1}, respectively (Tukey's HSD, $P < 0.05$). Similarly, the farm-scale income increased significantly ($P < 0.001$) for organic plant biomass, and it was most pronounced for Mucuna + Tithonia and sole Tithonia, followed by sole Mucuna and mineral fertilizer, as compared to the control (Tukey's HSD, $P < 0.05$). A positive Spearman Rank Correlation was recorded between the ISFM treatments and tomato yield or the farm-scale income ($r = 0.76$, $P < 0.05$), and between tomato yield and the farm- scale income ($r = 0.99$, $P < 0.05$). These results imply that smallholder farmers in SSA can effectively use the combination of Mucuna + Tithonia biomass materials or their sole applications as basal mulch to improve tomato production. Thus, these organic amendments could be an alternative and sustainable integrated soil fertility management strategy to boost tomato production and farm-scale income without jeopardizing the sustainability of the environment. However, this requires more efforts to adapt the different ISFM techniques to the specific needs of smallholder farmers, coupled with effective dissemination strategies that facilitate knowledge transfer and technology adoption.

KEYWORDS

Integrated Soil Fertility Management, Mucuna, Tithonia, Mineral Fertilizer, Tomato

1. INTRODUCTION

Sub-Saharan Africa (SSA) accounts for about 9% of global population with high food and nutrition insecurity that is partly due to poor and declining fertility status of the largely acid soils, with nitrogen and phosphorus as the most limiting mineral elements [1] [2]. Although mineral fertilizers are often used to correct soil acidity and improve the soil fertility status, SSA accounts for only 0.1% of global mineral fertilizer production and 1.8% of global mineral fertilizer use, with less than 10 kg ha^{-1} compared to 87 kg ha^{-1} for developed nations [3] [4]. Meanwhile, nutrient losses from arable fields are higher than the natural replenishment capacity of soils in SSA [5] [6]. The poor soil fertility status coupled with low mineral fertilizer inputs in SSA accounts for the low crop yields with huge gaps of over 30% between attainable potential and actual production [7] [8]. Besides exerting a multitude of deleterious effects on the environment, mineral fertilizers are expensive and unaffordable to many resource-poor smallholder farmers in SSA, which necessitates sustainable and affordable alternative soil amendment strategies that are adapted to the needs of smallholder farmers.

The concept of integrated soil fertility management (ISFM) is widely encouraged in SSA, and it is largely considered as a set of agronomic practices adapted to local conditions to maximize nutrient use efficiency and improve crop productivity. The ISFM techniques incorporate mineral fertilizer and locally available inputs (i.e. lime and rock phosphate) with organic matter (i.e. crop residues and compost) to replenish soil nutrients, improve agronomic efficiency and crop production [9]. Meanwhile, wood ash has been used as additives to mineral fertilizer and compost to improve quality and performances [10] [11]. Ash and ash-derived composts have high liming potential and reportedly improved soil physical and biochemical properties [12] - [15]. However, the reliance on plant materials for soil restoration partly depends on the availability and quality of plant residues. Tithonia diversifolia has high biomass and nutrient contents with 3.5% nitrogen, 0.37% phosphorus and 4.1% potassium [16] [17], but contains a few low recalcitrant compounds with 6.5% lignin and 1.6% polyphenol [18]. Tithonia demonstrates strong potential for soil rejuvenation [19] [20], and plant health management due to the presence of sesquiterpene lactones (tagitinins-terpene) and other antimicrobial substances that prevent pests and diseases [13]. Mucuna spp. has high N-fixing ability and abundant biomass for use in soil rejuvenation [21], and contains up to 3% nitrogen, 0.2% phosphorus and 1.4% potassium [22] - [24]. Mucuna exhibits antimicrobial and fauna properties, influences abundance and diversity of soil bacteria and fungi, as well as suppresses nematode populations [25] - [27].

Despite the agronomic importance of different ISFM strategies, their adoption by smallholder farmers has been relatively slow due to a combination of factors, which include poor understanding of the economic benefits, high labour demand and poor technology promotion strategies. Furthermore, most ISFM studies have mainly focused on the direct impact of ISFM on soil restoration parameters rather than the crop productivity and potential economic benefits for farmers. Although many studies have been conducted on the importance of mineral fertilizer and organic amendments as well as on their combinations in various ISFM strategies, few studies have been devoted to the potential benefits of combining different plant biomass materials in ISFM. To the best of our knowledge, there is no specific investigation on the potential benefits of combining Mucuna and Tithonia biomass materials as basal mulch for integrated soil fertility management in horticultural systems. Hence, this participatory ISFM field trial was intended to adapt and evaluate the potential of combining two plant biomass materials (Mucuna + Tithonia) as basal mulch to improve soil fertility and plant nutrition, so as to improve tomato yield and farm-scale income for smallholder farmers in Cameroon. Therefore, it is hypothesized that the interaction of organic mulches of Mucuna + Tithonia will improve tomato yields and farm-scale income as compared to sole Mucuna or Tithonia biomass applications and mineral fertilizer amendments or the control without any input.

2. MATERIALS AND METHODS

2.1. Experimental Site and Setup

This study was conducted on a smallholder farm in Lysoka-Buea, located at the foot of mount Cameroon, south- western Cameroon. The area is situated between latitudes 4˚3'N and 4˚12'N of the equator and longitudes 9˚12'E and 9˚20'E. The soils are mostly derived from volcanic rocks and generally fertile but have poor moisture retaining capacity [28]. The climate is characterized by a mono-modal rainfall and less pronounced dry season, with 85% - 90% relative humidity. Heavy rainfall events are between June and October while the dry season starts from November to May. The mean annual rainfall ranges from 2085 mm at Ekona that is on the leeward side of the mountain to 9086 mm at Debundscha that is on the windward side of the mountain [29]. Lysoka is located between Ekona and Debundscha and had 2875 mm annual rainfall between mid March and mid November [29]. The mean monthly temperature vary from 19°C to 30°C, while soil temperature at 10 cm depths decreases from 25°C to 15°C with increasing elevation from 200 m to 2200 m above sea level [30] [31].

The field site had been under intensive commercial banana production by the Cameroon Development Corporation (CDC) until 2009. The site was further used for smallholder intercropping system to cultivate subsistence crops like maize (Zea mays), (Manihot esculenta), okra (Abelmoschus esculentus), ginger (Zingiber officinale), beans (Phaseolus spp) and cowpeas (Vigna unguiculata) until 2013. In 2014, the site was cleared manually using cutlasses and partitioned into experimental plots of 20 m^2 (5 × 4 m) with a 1m buffer zone between plots. All the experimental plots were tilled manually using hoes and a cover crop (Mucuna cochinchinensis) was planted in March 2014, at 30 × 30 cm spacing and allowed to fallow for one year in order to establish a homogeneous soil. The Mucuna was harvested at maturity in March 2015, and Mucuna seeds and shells were separated from the biomass (leaves and stems). The biomass and seeds were sun-dried separately and preserved at room temperature for eventual use as basal mulch and propagation materials, respectively.

2.2. Tomato (Lycopersicon esculentum L.) Plants

Hybrid tomato (Lycopersicon esculentum L.) seeds (F1 Cobra 26; TECHNISEM® France) were purchased from a local agro-shop in Buea Cameroon. The F1 Cobra 26 is determinate with very good vigour, and combines tolerance/resistance to Tomato Yellow Leaf Curled Virus (YLCV) and Bacterial Wilt (Ralstonia solanacearum), which enable very good productivity, and viability to the Sahelian and tropical areas. It also has very good maturity with the first harvest from 65 days after planting, and produces square fruits of medium sizes (80 - 90 g) with uniform coloration. In addition, the fruits have good firmness that allows for good postharvest conservation.

The F1 Cobra 26 seeds were pre-germinated on a nursery bed of 2.5 × 1 m close to the experimental plots, which was prepared by clearing the site with a cutlass and the soil tilled manually using hoes. The tomato seeds were sown on the 26th of March 2015 on the nursery bed at an inter-row spacing of 15 × 15 cm. Vigorous tomato seedlings were transplanted from the nursery bed to the experimental plots on the 21st April 2015. The tomato seedlings were planted at a distance of 1 × 0.5 m on the experimental plots of 20 m^2 (5 × 4 m), with one plant per stand, giving a total of 35 stands per plot. In order to facilitate acclimatisation of tomato plants after transplanting, moisture was immediately provided for each plant via irrigation with tap water, while subsequent moisture was provided by rainfall. Each tomato plant received three litres of water, applied manually by pouring onto the plant-soil interface using watering cans. All 750 plants on the experimental site were staked manually with 1 m wooden sticks and the plants attached firmly to the stakes on the 15th of May 2015 using ropes.

2.3. Soil Fertility Amendments

The field trial was established as a randomized block design with four replicates per treatment. There were five treatment variables comprising a control with no input, mineral fertilizer, and two organic biomass mulches of Mucuna cochinchinensis and Tithonia diversifolia (Mexican sunflower), as well as their combination (Mucuna + Tithonia, at 1:1). Organic mulch materials were applied as single basal dose at the rate of 10 kg DW per plot (20 m^2) that is equivalent to 5 t ha^{-1} [32]. The organic mulches were evenly spread on the respective plots immediately after the tomato seedlings were transplanted, mulched and earthed-up three weeks later. The Mucuna biomass mulch was harvested from the same field site following one year of Mucuna cover crop fallow, while Tithonia biomass mulch was harvested from nearby roadsides and abandoned lands, dried in the sun for one week and stored at room temperature prior to field application. Labour for procuring Mucuna and Tithonia biomasses was provided by farmers who represented family labour according to common smallholder practice. Mineral fertilizer was applied by ring method at 5 cm from plants as two split doses of 90 kg ha^{-1} for each application. The first dose of granular NPK 20:10:10 (ADER®Cameroon) was applied on 14/05/2015 and immediately earthed-up with soil. The second dose of granular NPK 12:14:19 (ADER® Cameroon) was applied on the appropriate plots on 04/07/2015. Overall, a single dose application of 90 kg ha^{-1} fertilizer is approximately 87 kg ha^{-1} reported for developed nations [3] [4]. Considering the cost of 1 US Dollar or 500 FCFA per kg fertilizer, the total cost of mineral fertilizer was applied per hectare is 180 US Dollar or 90,000 FCFA.

2.4. Field Management

In order to facilitate technology appraisal and adoption, a participatory management approach was employed involving the local beneficiary smallholder farmers at all stages of production from field preparation and planting, to harvesting and marketing of produce. Apart from the initial irrigation performed immediately after transplanting of seedlings to the experimental plots, plant growth and performance depended entirely on rain-fed system based on the local rainfall regime. Meanwhile, management practices for weeds, pest and diseases were the same for all treatment plots. Before transplanting the tomato seedlings, the entire field was thoroughly weeded manually using cutlasses and hoes on 16/04/2015. After transplanting of the tomato seedlings, the field was monitored regularly for the emergence of weeds and when necessary, weeding was performed manually using hoes on 14/05/2015, 10/06/2015 and 02/07/2015. In order to maintain plant health and crop damage by pests and diseases within acceptable economic levels, frequent scouting was done for all plots to monitor the emergence of insect pests and diseases. When necessary, all plots were sprayed with appropriate doses of fungicide (Mancozan super; SCPA SIVEX International® France) or insecticides (Garmaline 80, AGROMAF® Cameroon; Cigogne 360, SCPA SIVEX International® France; and Acarius, SAVANA-Horizon Phyto Plus® Cameroon).

2.5. Harvest and Analyses

Since the main focus of this particular investigation was on the tomato yield and potential income gains for the resource-poor smallholder farmers, data on the soil dynamics were not considered in this study. A total of nine tomato harvests were performed within thirty-two days, starting on the 20th of June 2015 and ending on the 27th of July 2015. The tomato fruits were harvested twice a week from each treatment plot, counted and weighed separately using a top loading balance. The harvested tomato fruits were sold in the local markets and the market value was used to determine the farm-scale income generated by each treatment. The farm-scale income was calculated as the total income generated after computing the difference between the total tomato sales and the cost of inputs for the respective plots (i.e. farm-scale income = total tomato sales ? total cost of inputs). In order to evaluate the direct impact of ISFM treatments, we determined the additional tomato yields and farm-scale income derived from each ISFM treatment by calculating the difference between each ISFM treatment and the control (i.e. additional yield or income = ISFM treatment ? control). All treatment means were computed and the data sets were subjected to statistical analysis using STATISTICA 9.1 for Windows [33]. The effects of ISFM treatments (n = 5) on tomato performance (yield—total fresh fruit weight, number of fruits and the thousand fruit weight) and farm-scale income were assessed by analysis of variance (ANOVA), and significant means were compared by a post-hoc Tukey's HSD test ($P < 0.05$). Additionally, a nonparametric Spearman Rank Order Correlation ($P < 0.05$) was performed to measure the degree of association between the different treatments (n = 5) and tomato performance or farm-scale income.

3. RESULTS

The impact of ISFM techniques was mainly evaluated from the perspective of potential benefits for smallholder farmers (i.e. crop performance and economic gains) so as to provide greater insight that may facilitate ISFM technology adoption by resource-poor farmers in SSA, since technology adoption is based on knowledge and proven efficacy as well as potential economic gains.

3.1. Tomato Performance

Tomato performance was assessed as yield (fresh fruit weight) and the thousand fruit weight for each treatment. The average tomato yield measured as total fresh fruit weight ranged from 9.7 to 13.2 t ha^{-1} across the different treatments (Table 1). Tomato yield was significantly higher with the organic mulch amendments ($P < 0.001$), which is most pronounced for the interaction of Mucuna + Tithonia biomass and the sole Tithonia biomass application, followed by the sole Mucuna biomass and mineral fertilizer treatments, as compared to the control (Tukey's HSD, P <

0.05;Table 1). In comparison to the control, an additional 3.5 and 3.4 t ha^{-1} tomato yield was recorded for the interaction of Mucuna + Tithonia biomass and the sole Tithonia treatments, respectively. This additional increase is more than twice the additional amount recorded for the sole Mucuna and mineral fertilizer treatments with 1.8 and 1.5 t ha^{-1} (Figure 1). Furthermore, a positive Spearman Rank Correlation was recorded between the different ISFM treatments and tomato yields ($r = 0.76$, $P < 0.05$). Meanwhile, there was no significant difference or correlation between the different ISFM treatments and the thousand fruit weight.

3.2. Impact of ISFM on Farm-Scale Income

The potential for better crop performance and greater economic gains largely drive the interest of smallholder farmers in adopting new technologies against traditional practices. The total farm-scale income ranges between 7223 and 8520 US Dollar (3,128,663 and 4,260,136 FCFA) across the different treatments. The total income was significantly higher for the organic mulches ($P < 0.001$), and most pronounced for the interaction of Mucuna + Tithonia biomass and the sole Tithonia treatments, followed by the sole Mucuna biomass and mineral fertilizer treatments as compared to the control (Tukey's HSD, $P < 0.05$; Table 2). Correspondingly, the interaction of Mucuna + Tithonia and sole Tithonia biomass generated the highest farm-scale income, followed by the sole Mucuna biomass and mineral fertilizer treatments, in comparison to the control ($P < 0.001$; Figure 2). Furthermore, a significant positive Spearman Rank Correlation was recorded between the tomato yield and farm-scale income ($r = 0.99$, $P < 0.05$) and between the different treatments and farm-scale income ($r = 0.76$, $P < 0.05$).

4. DISCUSSION

According to The Economist [34], feeding nine billion people on earth by 2050 will require gains from narrowing the gap between the worst and best agricultural producers by taking advantage of new technologies. This view is largely supported by technological advancements in agriculture and highlights the need for sustainable integrated soil fertility management strategies. This is particularly important for smallholder farming systems that produce the majority of food that is consume in the low and middle-income countries. Correspondingly, the results of this study demonstrate the potential for ISFM to significantly narrow the gap between potential and actual tomato yields and income for smallholder farmers in SSA.

Table 1. The impact of different integrated soil fertility management treatments on tomato performance—yield and thousand fruit weight (mean ± STD); Data within column with different letters are significantly different according to Tukey's HSD, P < 0.05.

	Tomato performance	
Treatments	Yield [t ha^{-1}]	1000 Fruit weight
Control	9.7 ± 0.3 c	68.6 ± 16.6 a
Mineral fertilizer	11.2 ± 0.5 b	68.3 ± 16.6 a
Mucuna	11.5 ± 0.4 b	76.1 ± 16.4 a
Tithonia	13.1 ± 1.2 a	70.4 ± 10.1 a
Mucuna + *Tithonia*	13.2 ± 0.2 a	68.2 ± 10.0 a

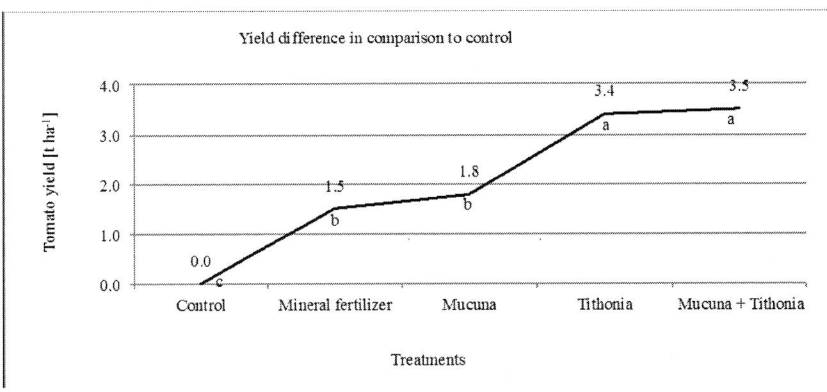

Figure 1. Additional tomato yield derived from the different integrated soil fertility management treatments in relation to the control; Data with different letters are significantly different according to Tukey's HSD, P < 0.05.

4.1. Impact of ISFM on Tomato Performance

Crop yield is determined by a combination of environmental factors and farm management practices that in turn influence soil fertility and plant nutrition. Despite applying 180 kg ha^{-1} mineral fertilizer that is more than twice the 87 kg ha^{-1} applied in developed nations [3] [4], organic biomass was better than mineral fertilizer that was only better than the control. The increased tomato yield recorded for Mucuna and Tithonia treatments is consistent with the results reported in other studies that attributed greater crop yields to improved soil nutrients from organic matter and adequate soil moisture, bulk density and temperature [17] [32]. Hence, the increased tomato yields with organic amendments can be attributed to increased soil nutrient availability and improved plant nutrient uptake resulting from better soil physical, chemical and biological properties [35]. Mulching with plant materials increases yields by stimulating decomposition and mineralisation in the rhizosphere, which improves soil nutrient availability, organic matter content, structure, moisture and temperature [36] [37]. In addition, differences in the quality of plant materials, nutrient content, texture, rate of decomposition, growth and vegetative matter turn over, availability and cost are important for effective ISFM [38] [39].

Table 2. Impact of integrated soil fertility management treatments on farm-scale income of smallholder farmers (expressed both in US Dollar and FCFA—local currency of Cameroon); Data within the same column with different letters are significantly different according to Tukey's HSD, $P < 0.05$.

	Farm-scale income	
Treatments	FCFA	US Dollar
Control	3128663 c	6257 c
Mineral fertilizer	3611854 b	7224 b
Mucuna	3688359 b	7377 b
Tithonia	4227923 a	8456 a
Mucuna + Tithonia	4260136 a	8520 a

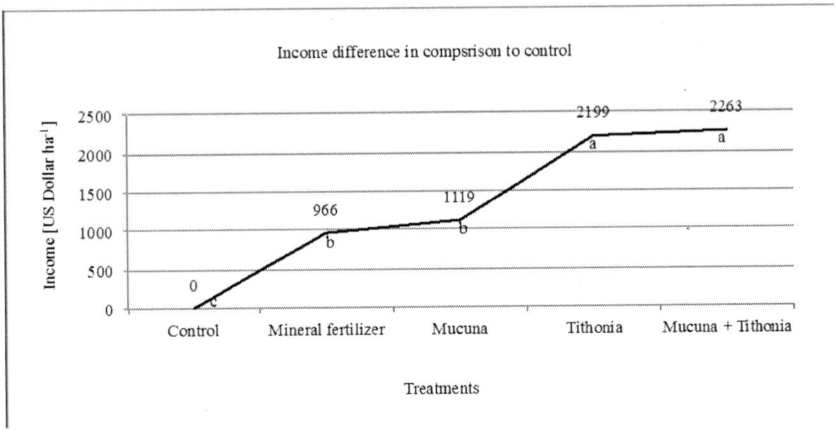

Figure 2. Additional income (US Dollars) derived from integrated soil fertility management treatments in relation to the control; Data with different letters are significantly different according to Tukey's HSD, $P < 0.05$.

Besides the influence of Mucuna and Tithonia on soil properties, they possibly enhanced both root and shoot physiological and morphological developments, which in turn increased productivity. This is consistent with greater root production and lifespan induced by compost and mulching as compared to mineral fertilizer [40]. In addition, mineral fertilizer induces high nitrate (NO_3^-) pulses, which could easily be leached as a result of its high mobility in the soil, coupled with the heavy rainfall condition of the study area. However, organic inputs would have released nitrogen more slowly, which enabled regular supply of nitrogen ions (i.e. NH_4^+ and NO_3^-) during a longer period of time to the root system that enhanced branching of axial roots and elongation of lateral roots, leading greater nutrient acquisition [41]. Furthermore, decomposing Mucuna and Tithonia biomass likely increased humic substances in the rhizosphere that modified root morphology and induced proliferation of lateral roots and root hairs, as well as increased the rate of differentiation for root cells [42] [43]. Hence, greater tomato yields recorded for Mucuna and Tithonia amendments could be due to the interaction of improved soil properties (i.e. physical, chemical and biological) and plant development (i.e. root physiology and morphology), which resulted in improved plant nutrition. The lack of significant difference or correlation between the ISFM treatments and tomato fresh weight could be attributed to the fact that some of the potential differences were compensated by water absorption from the soil that supplies all the water that the plant needs. It is suggested that comparison should be done on dry weight basis, which was not applicable in this case as the consumers preferably purchased fresh tomato fruits still having the original water content.

Although ISFM considers combining appropriate micro doses of mineral fertilizer either solely or in combination with locally available organic materials as an important strategy for soil fertility improvement, the use of different organic inputs either solely or combined would be a more sustainable ISFM alternative. Accordingly, the best tomato yield was recorded with the combined biomass of Mucuna + Tithonia and the sole Tithonia biomass application as compared to sole Mucuna, mineral fertilizer or the control. This is consistent with [44] who demonstrated advantages of Tithonia + Poultry manure compost (4:1) and sole Tithonia for improving soil properties and tomato performance (i.e. fruit yield and diameter, stem circumference, plant height, leaf area, number of branches and taproot length). The advantage of combining organic inputs is because organic materials differ in their rates of biodegradation. Some organic materials have faster decomposition and mineralization rates that are comparable with the fast rate of nutrient release by mineral fertilizers. Such organic inputs can be associated with materials that have slow rate of nutrient release to produce an effective amendment that is comparable or better than mineral fertilizers. Apparently, Mucuna and Tithonia biomasses provided fast nutrient release that enabled better tomato performance than mineral fertilizer throughout the plants' life cycle, leading to greater production. The higher tomato yields under combined organic biomass of Mucuna + Tithonia amendment that is comparable to the sole Tithonia biomass application supports our hypothesis of better performance with sustainable ISFM technologies comprising mainly a combination of different organic plant materials that advocates the use of mainly organic biomass in ISFM. Therefore, such fast nutrient release organic biomass materials can be integrated with other organic materials that release nutrients slowly, so as to enable regular and enhanced nutrient supply during the entire crop cycle, leading to improved plant nutrition and greater crop yield.

4.2. Impact of ISFM on Farm-Scale Income

The adoption of integrated soil fertility management technologies largely depends on the confidence of smallholder farmers towards these techniques, which is mainly determined by the crop production and income gained. However, inadequate knowledge on the best-bet ISFM options for specific crops and lack of confidence in the economic returns induces laxity of adopting such technologies by smallholder farmers. Thus, smallholder farmers in SSA have not experienced a significant decrease in yield gap between the actual and expected production per unit area of cultivable land. This is partly due to inadequate supply of the required inputs and high cost, poor technology adoption and lack of knowledge on potential benefits [45] [46]. Meanwhile, knowledge on the economic benefits of ISFM and effective technology dissemination is important for successful adoption. This study demonstrated the potential benefits of combining different organic biomass materials for improving the net productivity and financial gain for smallholder farmers. This strongly supports our hypothesis on better farm-scale gains with ISFM involving a combination of organic plant materials of Mucuna + Tithonia as compared to sole Mucuna biomass or mineral fertilizer applications and the control. Considering that Mucuna and Tithonia biomass used for this study were obtained via smallholder labour without extra financial cost, any additional yield and economic gains is direct benefit for smallholder farmers who mainly depend on family labour. Interestingly, compared to the current monthly minimum wage of about US Dollar 73 (36,270 FCFA) in Cameroon, the additional income derived from ISFM strategies in this study was more than 100%, ranging from US Dollar 153 to 1297 (76,505 to 648,281 FCFA) for organic biomass as compared to mineral fertilizer. Meanwhile, the opportunity cost for resource-poor farmers with little access to money is often 100% of the actual value due to other high priority uses of available funds and other investment opportunities [47]. Therefore, ISFM options with no extra financial costs like the Mucuna and Tithonia biomass materials used for this study are the best-bet soil fertility amendment options for the resource-poor smallholder farmers in SSA.

4.3. Mitigating Constraints of ISFM in Smallholder Systems

A plethora of views, paradigms and concepts on sustainable agriculture and natural resource management have arisen in recent decades, which are intended to boost agricultural productivity and provide sustainable solutions to one of the earth's most pressing challenges of preserving the natural resource base. These include integrated soil fertility management, conservation agriculture, integrated nutrient management, organic agriculture, and integrated natural resource management [48] [49]. However, all the diverse concepts and novel titles praise the potential merits of combining different techniques and promoting a combination of different agricultural technologies that are not mutually exclusive or completely overlapping [49] [50]. A major constraint for the use of organic materials in ISFM such as Mucuna and Tithonia biomass is the huge amount required and the labour cost required for collection, transportation and application [51] [52]. However, we obtained better economic returns for sole Tithonia or Mucuna + Tithonia compared to the control or fertilizer treatments, while Mucuna was similar to fertilizer treatment. Hence, Mucuna and Tithonia biomass are sustainable ISFM alternatives in smallholder farming systems because of their economic returns, coupled with their abundance and adaptability to various environments, rapid growth rate and high vegetative matter turnover [16] [53]. In addition, smallholder farmers can easily cultivate or harvest Mucuna and Tithonia biomass from abandoned sites or fallow land for use in their fields as soil amendments without any additional cost. It is also possible to plant Tithonia along the borders of farmers' fields to serve a dual purpose of land demarcation and for use in soil fertility management. By combining Mucuna and Tithonia biomasses, farmers can reduce the difficulties of obtaining a particular material by using reduced amounts in combination with other biomass materials that are relatively abundant. There by achieving comparable soil improvement that leads to greater crop yield and farm-scale income.

5. CONCLUSION

Besides the proven efficacy for Mucuna and Tithonia biomass for independently improving soil physical, chemical and biological properties, the Mucuna + Tithonia biomass combination was comparable to sole

Tithonia, which was better than sole Mucuna biomass that displayed a comparable performance with mineral fertilizer in terms of tomato yield and farm-scale gains. This makes Mucuna + Tithonia biomass a viable and important ISFM alternative for sustainable soil fertility management that is commensurate with the existing plethora of views, paradigms and concepts on sustainable agriculture and natural resource management. Hence, smallholder farmers can effectively adopt Mucuna and Tithonia biomass applied in combination or solely, to improve net productivity and farm-scale income without any negative externalities. However, this requires more efforts to adapt different techniques to the specific needs of smallholder farmers, with effective dissemination strategies that facilitate technology adoption. Although Tithonia is widely distributed on abandoned lands and roadsides in Cameroon, it can be planted on farm boundaries to enhance availability and reduce the associated labour and transport costs. Meanwhile, Mucuna biomass can be harvested from fallow sites where they are cultivated as green manure cover crops. Overall, this alternative soil fertility management technique has multiple advantages on soils including decontamination, mineralization, conservation, increased soil biota, erosion control and water conservation within the broader context of sustainable agriculture, and the specific concept of integrated soil fertility management in smallholder farming systems. Nonetheless, more in situ based studies are necessary to provide greater insights on the specific impacts of combining Mucuna and Tithonia biomass materials on the dynamics of soil biota (i.e. microbial abundance and diversity, food web ecology or, mycorrhiza dynamics, etc.). Thereby providing a holistic view of soil microorganism and integrated soil fertility management that improves soil fertility and plant nutrition, leading to greater crop productivity and income for farmers.

ACKNOWLEDGEMENTS

Funding for this study was provided by Research Grant of the Faculty of Agriculture and Veterinary Medicine— FAVM, University of Buea, and supported by the Research and Modernization Allowance of the Ministry of Higher Education of Cameroon. We are grateful to Ngu George, Bezeng Fidelis and Lyonga Efunganyi (FAVM, University of Buea) for their assistance with field management, and Asongwe Joseph "General" who secured the experimental site.

REFERENCES

1. Kadiata, B.D. and Lumpungu, K. (2003) Differential Phosphorus Uptake and Use Efficiency among Selected Nitrogen-Fixing Tree Legumes over Time. Journal of Plant Nutrition, 26, 1009-1022.

2. Sanginga, N., Lyasse, O., Diels, J. and Merckx, R. (2003) Balanced Nutrient Management Systems for Cropping Systems in the Tropics: From Concept to Practice. Agriculture, Ecosystems, and Environment, 100, 99-102.

3. Bationo, A., Hartemink, A., Lungu, O., Naimi, M., Okoth, P., Smaling, E. and Thiombiano, L. (2006) African Soils: Their Productivity and Profitability of Fertilizer Use. Proceedings of the African Fertilizer Summit, Abuja, 9-13 June 29.

4. FAO (2007) Policies and Actions to Stimulate Private Sector Fertilizer Marketing in Sub-Saharan Africa. Agricultural Management, Marketing and Finance Occasional Paper 15.

5. Smaling, E.M.A. and Braun, A.R. (1996) Soil Fertility Research in Sub-Saharan Africa: New Dimensions, New Challenges. Communication in Soil Science, 7, 365-386.

6. Sanchez, P. and Jama, B. (2002) Soil Fertility Replenishment Takes off in East and Southern Africa. In: Vanlauwe, B., Diels, J., Sanginga, N. and Merckx, R., Eds., Integrated Plant Nutrient Management in Sub-Saharan Africa: From Concept to Practice, CABI, Wallingford, 23-46.

7. Sanchez, P.A. (2002) Soil Fertility and Hunger in Africa. Science, 129, 2019-2020.

8. Bekunda, B., Sanginga, N. and Woomer, P.L. (2010) Restoring Soil Fertility in Sub-Sahara Africa. Advances in Agronomy, 108, 184-236.

9. Vanlauwe, B., Bationo, A., Giller, K.E., Merckx, R., Mokwunye, U., Ohiokpehai, O., Pypers, P., Tabo, R., Shepherd, K.D., Smaling, E.M.A., Woomer, P.L. and Sanginga, N. (2010) Integrated Soil Fertility Management. Operational Definition and Consequences for Implementation and Dissemination. Outlook on Agriculture, 39, 17-24.

10. Kuba, T., Tsch&oumllll, A., Partl, C., Meyer, K. and Insam, H. (2008) Wood Ash Admixture to Organic Wastes Improves Compost and Its Performance. Agriculture, Ecosystem and Environment, 127, 43-49.

11. Bougnom, B.P. and Insam, H. (2009) Ash Additives to Compost Affect Soil Microbial Communities and Apple Seedling Growth. Die Bodenkultur, 60, 9-19.

12. Noble, A.D., Zenneck, I. and Randall, P.J. (1996) Leaf Litter Ash Alkalinity and Neutralisation of Soil Acidity. Plant and Soil, 179, 293-302.

13. Adoyo, F., Mukalam, J.B. and Enyola, M. (1999) Using Tithonia Concoctions for Termite Control in Busia District, Kenya. ILEIA Newsletter, 13, 24-25.

14. Bougnom, B.P., Mair, J., Etoa, F.X. and Insam, H. (2009) Composts with Wood Ash Addition: A Risk or a Chance for Ameliorating Acid Tropical Soils? Geoderma, 153, 402-407.

15. Bougnom, B.P., Knapp, B.A., Elhottová, D., Koubovác, A., Etoa, F.X. and Insam, H. (2010) Designer Compost with Biomass Ashes for Ameliorating Acid Tropical Soils: Effects on the Soil Microbiota. Applied Soil Ecology, 45, 319-324.

16. Olabode, O.S., Sola, O., Akanbi, W.B., Adesina, G.O. and Babajide, P.A. (2007) Evaluation of Tithonia diversifolia (Hemsl.) A Gray for Soil Improvement. World Journal of Agricultural Sciences, 3, 503-507.

17. Agbede, T.M. and Afolabi, L.A. (2014) Soil Fertility Improvement Potentials of Mexican Sunflower (Tithonia diversifolia) and Siam Weed (Chromolaena odorata) Using Okra as Test Crop. Archives of Applied Science Research, 6, 42-47.

18. Jama, B., Palm, C.A., Buresh, R.J., Niang, A., Gachengo, C. and Nziguheba, G. (2000) Tithonia diversifolia as a Green Manure for Soil Fertility Improvement in Western Kenya: A Review. Agroforestry Systems, 49, 201-221.

19. Ojeniyi, S.O., Odedina, S.A. and Agbede, T.M. (2012) Soil Productivity Improving Attributes of Mexican Sunflower (Tithonia diversifolia) and Siam Weed (Chromolaena odorata). Emirates Journal of Food and Agriculture, 24, 243-247.

20. Agbede, T.M., Adekiya, A.O. and Ogeh, J.S. (2014) Response of Soil Properties and Yam Yieldto Chromolaena odorata (Asteraceae) and Tithonia diversifolia (Asteraceae) Mulches. Archives of Agronomy and Soil Science, 60, 209-224.

21. Mathews, J., Joseph, K., Lakshmanan, R., Jose, G., Kothandaraman, R. and Jacob, C.K. (2003) Effect of Bradyrhizobium Inoculation on Mucuna bracteata and Its Impact on the Properties of Soil under Hevea. Proceedings of the 6th International PGPR Workshop, 5-10 October 2003, Calicut, 29-33.

22. Mathews, J. and Leong, T.T. (2000) Performance of Two New Legume Species in Oil Palm Planting. In: Pushparajah, E., Ed., Proceedings of the International Planters Conference on Plantation Tree Crops in the New Millenium: The Way Ahead, Volume 1, The Incorporated Society of Planters, Kuala Lumpur, 325-339.

23. Shaharudin, B. and Yow, T.K. (2000) Establishment of Leguminous Cover Plant (Mucuna bracteata). Poster Presentation, The Incorporated Society of Planters, Kuala Lumpur, 17-20 May 2000, 317-323.

24. Chiu, S.B. and Bisad, M. (2006) Mucuna bracteata—Biomass, Litter and Nutrient Production. The Planter, 82, 247-254.

25. Vargas-Ayala, R., Rodríguez-Kábana, R., Morgan-Jones, G., McInroy, J.A. and Kloepper, J.W. (2000) Shifts in Soil Microflora Induced by Velvetbean (Mucuna deeringiana) in Cropping Systems to Control Root-Knot Nematodes. Biological Control, 17, 11-22.

26. Rayavarapu, A.K. and Kaladhar, D.S.V.G.K. (2011) Evaluation of Antimicrobial Activity of Mucuna pruriens on Plant Pathogens. Asian Journal Biochemical and Pharmaceutical Research, 2, 593-600.

27. Pujari, S.A. and Gandhi, M.B. (2013) Studies on Effects of Seed and Leaf Extracts of Mucuna pruriens on Some Common Bacterial Pathogens. Journal of Environmental Research and Development, 8, 50-54.

28. Cheek, M. (1992) A Botanical Inventory of Mabeta-Moliwe Forest. Report to ODA/MCP, Royal Botanic Gardens, Kew, 122.

29. Fraser, P., Banks, H., Brodie, M., Cheek, M., Daroson, S., Healey, J., Marsden, J., Ndam, N., Nning, J. and McRobb, A. (1999) Plant

Succession on the 1922 Lava Flow of Mt. Cameroon. In: Timberlake, J. and Kativu, S., Eds., African Plants: Biodiversity, Taxonomy and Uses, Royal Botanic Garden, Kew, 253-262.

30. Payton, R.W. (1993) Ecology, Altitudinal Zonation and Conservation of Tropical Rainforest of Mount Cameroon. Final Project-Report R4600, ODA, London.

31. Fraser, P.J., Hall, J.B. and Healing, J.R. (1998) Climate of the Mount Cameroon Region, Long and Medium Term Rainfall, Temperature and Sunshine Data. School of Agricultural and Forest Sciences, University of Wales Bangor, MCP-LBG, Limbe, 56 p.

32. Agbede, T.M., Adekiya, A.O. and Ogeh, J.S. (2013) Effects of Chromolena and Tithonia Mulches on Soil Properties, Leaf Nutrient Composition, Growth and Yam Yield. West African Journal of Applied Ecology, 21, 15-29.

33. StatSoтt (2010) STATISTICA 9. 1 for Windows. StatSoft Inc., Tusla.

34. The Economist (2011) The 9 Billion-People Question: A Special Report on Feeding the World. February 26, 2011.

35. Kolawole, O.K., Awodun, M.A. and Ojeniyi, S.O. (2014) Soil Fertility Improvement by Tithonia diversifolia (Hemsl.) A Gray and Its Effect on Cassava Performance and Yield. International Journal of Engineering and Science, 3, 36-43.

36. Opara-Nadi, O. (1993) Effect of Elephant Grass and Plastic Mulches on Soil Properties and Cowpea Yield on an Ultisol in Southeastern Nigeria. In: Mulongoy, K. and Merckx, R., Eds., Soil Organic Matter Dynamic and Sustainability of Tropical Agriculture, John Wiley and Sons, Chichester, 351-360.

37. Ojeniyi, S.O. and Adetoro, A.O. (1993) Use of Chromolaena Mulch to Improve Yield of Late Season Tomato. Nigerian Journal of Technical Education, 10, 144-149.

38. Kothandaraman, R., Mathew, J., Krishnakumar, A.K., Joseph, K., Jayarathnam, K. and Sethuraj, M.R.M. (1989) Comparative Efficiency of Mucuna bracteata D.C. and Peuraria phaseoloides Benth. On Soil

Nutrient Enrichment, Microbial Population and Growth of Hevea. Indian Journal of Natural Rubber Research, 2, 147-150.

39. Awodun, M.A. and Ojeniyi, S.O. (1999) Use of Weed Mulches for Improving Soil Fertility and Maize Performance. Applied Tropical Agriculture, 2, 26-30.

40. Baldi, E., Toselli, M., Eissenstat, D.M. and Marangoni, B. (2010) Organic Fertilization Leads to Increased Peach Root Production and Lifespan. Tree Physiology, 30, 1373-1382.

41. Boukcim, H., Pages, L. and Mousain, D. (2006) Local or Supply Modifies the Root System Architecture of Cedrus atlantica Seedlings Grown in a Split Root Device. Journal of Plant Physiology, 163, 1293-1304.

42. Concheri, G., Nardi, S., Reniero, F. and Dell' Agnola, G. (1996) The Effects of Humic Substances within the Ah Horizon of a Calcic Luvisol on Morphological Changes Related to Invertase and Peroxidase Activities in Wheat Roots. Plant Soil, 179, 65-72.

43. Canellas, L.P., Olivares, F.L., Okorokova-Facanha, A.L. and Facanha, A.R. (2002) Humic Acids Isolated from Earthworm Compost Enhance Root Elongation, Lateral Root Emergence, and Plasma Membrane H+-ATPase Activity in Maize Roots. Plant Physiololgy, 130, 1951-1957.

44. Babajide, P.A., Olabode, O.S., Akanbi, W.B., Olatunji, O.O. and Ewetola, E.A. (2008) Influence of Composted Tithonia Biomass and N-Mineral Fertilizer on Soil Physico-Chemical Properties and Performance of Tomato (Lycopersicon lycopersicum). Research Journal of Agronomy, 2, 101-106.

45. Vlek, P.L.G. (1990) The Role of Fertilizers in Sustaining Agriculture in Sub-Saharan Africa. Fertilizer Research, 26, 327-339.

46. Heisey, P.W. and Mwangi, W. (1996) Fertilizer Use and Maize Production in Sub-Saharan Africa. CIMMYT Economics Working Paper 96-01, CIMMYT, Mexico.

47. CIMMYT (1988) From Agronomic Data to Farmer Recommendations. An Economics Training Manual. Completely revised Edition, Mexico DF.

48. Lee, D.R. (2005) Agricultural Sustainability and Technology Adoption: Issues and Policies for Developing Countries. American Journal of Agricultural Economics, 87, 1325-1334.

49. Rosegrant, M.W., Koo, J., Cenacchi, N., Ringler, C., Robertson, R., Fisher, M., Cox, C., Garrett, K., Perez, N.C. and Sabbagh, P. (2014) Food Security in a World of Natural Resource Scarcity: The Role of Agricultural Technologies. International Food Policy Research Institute, Washington DC, 154 p.

50. Knowler, D. and Bradshaw, B. (2007) Farmers' Adoption of Conservation Agriculture: A Review and Synthesis of Recent Research. Food Policy, 32, 25-48.

51. Meertens, H.C.C. (2003) The Prospects for Integrated Nutrient Management for Sustainable Rainfed Lowland Rice Production in Sukumaland, Tanzania. Nutrient Cycling in Agroecosystem, 65, 163-171.

52. Chianu, J.N. and Tsujii, H. (2005) Integrated Nutrient Management in the Farming Systems of the Savannas of Northern Nigeria: What Future? Outlook Agriculture, 34, 197-202.

53. Tanimu, J., Lyocks, S.W.J. and Tanimu, Y. (2012) Growth and Biomass Production Assessment of Some Herbaceous Legumes for Soil Conservation in Nigeria. Journal of Occupational Safety and Environmental Health, 1, 38-44.

Chapter 4

Agroforestry and the Improvement of Soil Fertility: A View from Amazonia

Rachel C. Pinho,[1] Robert P. Miller,[2] and Sonia S. Alfaia[1]

[1]National Institute for Research in Amazon, INPA/CPCA, Agrarian Sciences Research Center, Avenue André Araújo, 69038-000 Manaus, AM, Brazil

[2]United Nations Development Program, (PNUD), Cj SHIS QI 25 Cj 03 C, 71640-220 Brasília, DF, Brazil

ABSTRACT

This paper discusses the effects of trees on soil fertility, with a focus on agricultural systems in Amazonia. Relevant literature concerning the effects of trees on soil physical and chemical properties in tropical, subtropical, and temperate regions is reviewed, covering both natural ecosystems and agroecosystems. Soil carbon, in the form of organic matter, is considered as an indicator of biological activity as well as in relation to policy issues such as carbon sequestration and climate change. In the case of tropical soils and Amazonia, information on the effects of trees on soils is discussed in the context of traditional agriculture systems, as well as in regard to the development of more sustainable agricultural alternatives for the region. Lastly, attention is given to a case study in the savanna region of Roraima, northern Brazil, where a chronosequence of indigenous homegarden agroforestry systems showed clear effects of management practices involving trees on soil fertility. The use of diverse tree species and other practices employed in agroforestry systems can represent alternative forms of increasing soil fertility and maintaining agricultural production, with important practical applications for the sustainability of tropical agriculture.

1. INTRODUCTION

According to a study by the World Agroforestry Centre, ICRAF, 43% of the planet's agricultural lands (more than a billion hectares) has more than 10% tree cover [1]. A lesser but still significant area of agricultural land, 160 million hectares, has more than 50% tree cover. The potential of trees to bring improvements in nutrition, income, housing, health, energy needs, and environmental sustainability in the agricultural landscape has guided ICRAF's mission, with the presence of trees being the principal component of an "evergreen agriculture" [2]. Within the array of benefits brought by trees, an important element is the positive effect of trees on soil properties and consequently benefits for crops. This paper explores current knowledge as to this relation between trees and soil, based on agroforestry systems research, as well as studies innon-agricultural or natural environments that demonstrate effects of trees on soil. Although we consider information from various ecosystems and biomes, the focus will be on Amazonia, where the authors have most of their experience. This focus on Amazonia is also due to the strong policy demands for the development of more sustainable agricultural systems in the region, as alternatives to forms of land use that have shown significant and negative impacts on natural resources and ecosystem services, such as deforestation for extensive cattle ranching. In this scenario, agroforestry systems have been indicated as one of the more promising alternatives to achieve a more sustainable agriculture, in greater equilibrium with the environment [3].

The presence of trees in farming systems, although an ancient practice, began to gain institutional attention during the 1970s and 1980s, with the beginning of studies on "agroforestry systems". One of the principal definitions employed in this context was that proposed by Lundgren and Raintree in 1982: "Agroforestry is a collective name for land-use systems and technologies where woody perennials (trees, shrubs, palms, bamboos, etc.) are deliberately used on the same land-management units as agricultural crops and/or animals, in some form of spatial arrangement or temporal sequence. In agroforestry systems there are both ecological and economical interactions between the different components" [4] (Figures 1, 2, and 3).

Figure 1. Agroforestry system with rubber (Hevea brasiliensis), cacao (Theobroma cacao), and açaí (Euterpe oleracea) in Tomé-açu, Pará, showing the litter layer that is typically found in such multistrata systems.

While trees in general can provide a number of environmental benefits in both rural and urban landscapes, and play key roles in ecosystem services provided by natural areas, in this paper we will restrict our focus to the effects of trees on soil fertility, in the specific context of agricultural systems. Although the benefits that trees can provide on rural properties such as food security, household income, economic stability, and thermal comfort (shade) are most often associated with their products, such as fruit, timber, or other items, the inclusion of trees in agricultural systems can also optimize nutrient cycling and have positive effects on soil chemical and physical properties. This process is especially important in tropical soils, where a high degree of weathering has created deep, leached soils that are poor in plant nutrients [5, 6]. Although poor in nutrients, tropical soils are very rich in biodiversity, with higher diversity and biomass of microorganisms than temperate soils, with these being the principal agents mediating the supply of nutrients to the soil by means of the decomposition of organic matter, derived from the vegetation [7–9].

Figure 2. Agroforestry system in initial phase, with black pepper (Piper nigrum) as principal cash crop, interplanted with cupuaçu (Theobroma grandiflorum) and açaí (Euterpe oleracea) for future fruit production, as well as timber trees (mahogany—Swietenia macrophylla and ipê—Tabebuia sp.) and Brazil nut (Bertholletia excelsa) as long-term products.

In the humid tropics, the removal of surface litter or organic matter generally results in the depletion of soil fertility in a few years [10, 11]. In agricultural systems practiced by traditional peoples, this limitation is circumvented by using the land for a short period (generally 2-3 years), after which the cultivated areas are left to fallow with natural regeneration of secondary vegetation. The associated ecological interactions reestablish nutrient cycling and recuperate soil qualities, after which the area can once again be used for agriculture [12,13]. This is the basis for shifting cultivation in Amazonia, a system that has permitted native populations to manage their natural resources over centuries, with small-scale environmental impacts that do not exceed the support capacity and resilience of ecosystems. However, the present-day situation of population growth and increasing pressure on agricultural lands lead to situations where there is demand for more intensive land use. This most often implies in repeated burning, the cheapest way to prepare land for planting, which can interrupt processes of nutrient cycling and accumulation, leading to loss of soil fertility and consequently slowing the recuperation of natural vegetation during fallow cycles [14].

Figure 3. Multistrata agroforestry system in Tomé-açu, Pará, with harvest of hogplum (Spondias mombin) grown as the upper canopy over cacao and açaí.

In light of the present-day situation of Amazonia, where there are now good reasons and policy demands to balance conservation with development, it is necessary to think in terms of agricultural systems that optimize nutrient cycling and permit permanent or semipermanent production, as well as minimize dependence on external inputs and have low environmental impact. The inclusion of diverse tree species is a key element in maintaining the production of organic matter and generating other positive benefits, as well as allowing the diversification of products. However, before we discuss topics specific to tropical soils and Amazonia, the following sections will review general information about the influence of trees on soil and their role in accumulation of soil carbon stocks.

2. A GENERAL VIEW OF THE INFLUENCE OF TREES ON SOIL FERTILITY

One of the pioneer studies to measure the effects of individual trees on soils was that by Zinke [15], who looked at pines growing on dunes in northern California, USA. His study found that under trees, certain soil properties exhibited a pattern of radial symmetry, with changes in pH, nitrogen, cations, and cation exchange capacity varying according to distance from the tree trunk, with a peak in these characteristics at a certain distance.

Subsequent studies also demonstrated patterns in the variation of soil characteristics as influenced by trees, such as in tropical savannas [16, 17], deserts [18], and areas of temperate forests [19–23]. In analyzing soil characteristics under individual tree crowns in Kenyan savannas, Belsky et al. [16] found greater levels of mineralizable N, microbial biomass, P, K, and Ca underneath the crowns when compared to open savanna. Burke et al. [17] explain that in dry savannas the strong limitation on water availability permits only punctuated establishment of trees and shrubs but that under crowns cycling occurs in a different form than in open grasslands, with the possibility of soil enrichment in a scale of decades. However, such soil changes can be reverted with the death of the tree or by fires. Belsky et al. [16] also point out the effect of nutrients deposited in dung by birds and large mammals that utilize trees as resting places or roosts.

Such patterns form what have been called “islands of fertility” or “resource islands” created by trees or bushes, generally in savannas or desert areas.

The microenvironment of these "islands" can also influence the composition of the herb stratum [16, 19], soil density [19, 20], and earthworm activity [20, 23] among other factors, allowing the creation of positive feedbacks that favor plant establishment and productivity [23, 24]. At the same time, these patterns can be important indicators of stability or risk of desertification in such areas [17,18].

Studies of forests in temperate climates indicate variations in soil that can be related to individual tree species. Besides the expected correlations, such as greater levels of N under legumes [20] or lower pH under species that produce acidifying litter, such as Pinus spp. [20, 23], other interesting interactions show that different species can alter soil in distinct ways, with variations in the increment of soil carbon [20], exchangeable Ca and Mg and per cent base saturation [21, 23].

In a study of 14 tree species in Poland, Reich et al. [23] found varied effects on soil characteristics; however, these effects were significantly related to the level of Ca in litter, independent of the species. Trees producing litter rich in Ca were associated with soils with greater pH, exchangeable Ca, and per cent base saturation, as well as greater rates of forest floor turnover and greater diversity and abundance of earthworms. Dijkstra [22] emphasizes that the rate of mineralization of organic Ca is a fundamental factor in this process, since it determines the immediate availability of this nutrient in the soil and can vary between species.

The study of vertical patterns of the distribution of nutrients in soil can indicate other phenomena that are not detected when only the horizontal distribution of nutrients is examined. In an evaluation of more than 20,000 globally distributed soil profiles, the greater part in temperate climates, Jobbágy and Jackson [25] found that cycling mediated by plants exerts a marked influence on the vertical distribution of nutrients in the soil, especially in the case of more limiting nutrients such as P and K. Patterns of greater concentration of these nutrients in surface layers (0–20 cm) were attributed to the fact that since these are more important to plants, they are subject to greater uptake and cycling, being absorbed from deeper layers and returned to the soil surface through litterfall and rain water throughfall. This process of uptake functions in opposition to leaching, which moves nutrients downward and acts more strongly on those nutrients that are in lessdemand by plants. If a nutrient is not limiting, its movement in the soil profile will be more influenced by leaching than by cycling and it will present higher concentrations at greater depth, as occurs with Na, Cl, and Mg [25]. In Poland, Ulery et al. [20] found this sort of pattern in soils influenced by the presence of four planted tree species, with increments of almost 3 times as much K in the surface layer in relation to the original soil before planting, while below 20 cm this increment was absent or negative. Their study also showed a high degree of leaching of Na, which is less in demand by plants.

Associated with biological cycling and leaching, other processes that influence the vertical distribution of nutrients in soil are atmospheric deposition and weathering (Trudgill, 1988 as cited by [25]). However, atmospheric deposition is considered minor when compared to the annual uptake by plant communities and generally has little influence on the vertical distribution of nutrients [25, 26]. The degree of weathering, however, appears to have a marked influence on the vertical distribution of nutrients, such that in more weathered soils the pattern of concentration in the surface layer is accentuated [25]. This emphasizes the importance of biological cycling in supplying nutrients in weathered soils, as is found in the greater part of the tropics and will be discussed in a subsequent section.

3. TREES AND SOIL CARBON

Trees add organic matter to the soil system in various manners, whether in the form of roots or litterfall or as root exudates in the rhizosphere [27]. These additions are the chief substrate for a vast range of organisms involved in soil biological activity and interactions, with important effects on soil nutrients and fertility. In participating in these complex processes, trees contribute to carbon accumulation in soils, a topic that is increasingly present in discussions on the mitigation of greenhouse gases associated with global warming and climate change. Although carbon (C) constitutes almost 50% of the dry weight of branches and 30% of foliage, the greater part of C sequestration (around 2/3) occurs belowground, involving living biomass such as roots and other belowground plant parts, soil organisms, and C stored in various soil horizons [28].

In a study that gathered information from sites around the world, Nair et al. [29] found values for soil organic C stocks ranging from 6.9 to 302 Mg ha^{-1}. Despite the great amplitude of these values, attributed to the variation between systems, ecological regions, and soil types, the study revealed a general trend of increasing soil C sequestration in agroforestry when compared to other land-use practices, with the exception of forests.

Although the ability of soils to accumulate C is generally related to characteristics that are little influenced by management, such as texture (clay soils typically accumulate more C than sandy soils), some management practices can influence soil C sequestration, particularly the insertion of trees in agricultural systems. Soils in various sites studied by Takimoto et al. [30] in the African Sahel were not markedly different among each other in terms of their characteristics such as pH, bulk density, and particle size, such that variations in their C contents seemed to be related to the influence of trees. In 8-year-old alley-cropping systems with five different species, for example, the authors found that greater C content is nearer to the trees. However, the greater part of this C was found in the form of particles of size between 250–2000 μm, fractions that are considered to be large and less stable. In systems where trees where present for more than 30 years (parklands), there was a predominance of soil C in smaller fractions (<53 μm), which are more stable and thus represent a more "protected" form of C.

In homegardens in India, Saha et al. [31] found levels of soil organic C to be 30% and 114% greater than in coconut plantations and rice paddies, respectively, when homegardens where small (less than 0.4 ha) and levels 18% and 94% greater in larger homegardens (larger than 0.4 hectares). Although the authors do not elaborate on the reasons for decrease in soil C in the larger homegardens, this may be related to a drop off in the deposition of organic residues, which generally occurs in proximity to dwellings. In the Northeast of Spain, Howlett et al. [32] studied the levels of soil C in silvopastoral systems composed of different species, and found that systems with birch (Betula pendula) presented greater levels of soil C than systems with pine (Pinus radiata). This difference was attributed to the fact that P. radiata create an understory environment that is less conductive to plant growth by the formation of a slowly decomposing duff layer, which inhibits germination and growth of other species and, over time, reduces C inputs to the soil. Both silvopastoral systems presented levels of C similar to pastures without trees, and in the case of the more stable C fractions (<53 μm) in the upper layers (0–50 cm), the pasture without trees presented greater levels than the silvopastoral systems. However, the authors observed that these levels may not accurately represent long-term C sequestration potential, since pasture tillage led to more free (unoccluded) silt+clay-sized fraction recovery, which translated into more storage in the pasture as compared with the pine and birch stands. In other situations tillage is reported to cause significant loss of surface soil organic C and disruption of the process of macroaggregate formation, increasing bioavailable sources of C from smaller aggregates [28, 30–32]. Tonucci et al. [33] consider that on the short term, C levels in pasture soils can be greater than under systems with trees due to the faster turnover of the grass root system, as well as the greater bulk density of soils in pasture systems, that can lead to higher C values than found in less dense soils.

The lag inachieving positive results in terms of soil C accumulation under trees are similar to what has been observed in the transition to continuous no-till agriculture. According to Derpsch et al. [34], clear increases in soil organic matter only appear 5–10 years after the adoption of this form of cropping system. Climate conditions also affect C accumulation in soils, since temperature and humidity greatly affect the activity of microbial communities and the breakdown of organic matter [28]. Takimoto et al. [30], for example, attributed, at least partly, To the low levels of C found in their study of soils in the African Sahel to the high temperatures in that region.

4. TREES AND SOIL BIODIVERSITY

In Amazonia, the diversity and density of macroinvertebrates in the soils of natural forests are considered high, and the conversion of these ecosystems to other forms of land use can cause severe changes in their populations [35, 36]. Barros et al. [37] cite soil macroinvertebrates as sensitive indicators of natural systems, land use, and management, with agroforestry systems presenting a greater abundance and diversity when compared to other forms of agricultural land use. Tapia-Coral et al. [38] found greater densities and biomass of Diplopoda and Isopoda in the litter layer of agroforestry systems than in adjacent secondary forests. Secondary forests, however, presented greater values for Isoptera, principally during the rainy season. This is an indication that such systems can support sufficiently abundant and diverse communites of these organisms that play an important role in ecological processes such as breakdown and cycling of organic matter. However, agroforestry systems encompass a range of practices, such that there is considerable variation in the quality of organic matter produced and the effects of vegetation on soil humidity and temperature regime, with significant effects on the soil macrofauna community [37].

In a study to evaluate the occurrence and diversity of arbuscular mycorrhizal fungi in soil under different land uses in Amazonia (high forest, old secondary forest, young secondary forests, agroforestry systems, crops, and pasture), Leal et al. [39] found great variation in species richness among samples, independent of the land use system. On the other hand, in these same systems Jesus et al. [40] observed that the principal differences in the composition and structure of the bacterial community were related to alterations in soil attributes, which in turn are correlated to land use. Community structure changed significantly along gradients of cation saturation and pH. The similarity observed between the communities of primary and secondary forest indicates the recuperation of the bacterial community during the process of succession that occurs after the abandonment of areas used for agriculture, with the return of arboreal vegetation.

5. TREES ON TROPICAL SOILS

In contrast to temperate ecosystems, where soils generally are more fertile and a greater part of the nutrients is supplied by the weathering of parent material, conditions of high temperatures and rainfall in the tropics accelerate soil processes, including loss of nutrients, so that the greater stock of nutrients is found held in the biomass and made available through decomposition [5–7, 9]. Although there are exceptions to this rule, such as soils of more recent volcanic origin in Asia, or the floodplains of the Amazon River that annually receive sediments from the Andes, the majority of soils in the Amazon Basin are typical examples of low-fertility tropical soils, having been formed on ancient sediments. The high temperature and humidity of tropical climates are conducive to the decomposition of organic matter, so that there is not only the release of nutrients but also the formation of negatively charged particles, which help to retain cations such as K, Ca, and Mg and maintain them in constant interface with the soil solution, where they can be absorbed by plants. In tropical systems, therefore, a soil cover of organic matter is essential to maintain adequate conditions for the soil micro, meso, and macrofauna that carry out this cycling [41]. As such, many of the studies on the effects of trees on tropical soil concentrate on the importance of organic matter made available in great part through litterfall.

Litter production can be a very important contribution in systems where perennial crops such as cacao and coffee are grown under the shade of trees. Souza et al. [42] observed that litter production in a coffee plantation shaded with diverse tree species was similar to that of native forests in the same region. Araujo and Collier [43], working in an agroforestry system with four fruit trees in a transition region between savanna and Amazonian forest in Brazil, indicated the importance of litterfall in the dry season, when compared to conventional cropping systems. Jaramillo-Botero et al. [44], in analyzing the effects of native trees planted as shade for coffee in southeastern Brazil, found that the quantity of accumulated litter and level of K in the soil was positively influenced by the number of trees present in a distance of 0 to 3 meters from the coffee bushes. Since K is a very mobile element, K enrichment may be partly due to throughfall and stemflow, as observed by Pinho et al. [45] in coffee agroforestry systems in the same region. Also in coffee systems, Jesus et al. [46] found higher pH and base saturation in areas intercropped with rubber (Hevea brasiliensis) than in monocultures. These studies suggest that roots of trees occupy deeper soil layers that may not be accessible to other crops, or that they are more efficient in extracting nutrients, due to their greater size and biomass or to other factors such as mycorrhizal associations. As such, when compared to other agricultural systems in tropical conditions, agroforestry can accumulate greater amounts of carbon and can help maintain soil fertility through a more efficient cycling of nutrients and a reduction of losses through leaching and erosion.

In a number of situations, intercropping with legume species that fix atmospheric N means that this element will be provided through decomposition to other plants, a practice commonly known as "green manuring" [47,48]. Mochiutti and Queiroz [49], working in the state of Amapá, eastern Amazonia, found that N levels were two times higher in the litter under improved fallows planted with the native legume taxi-branco (Sclerolobium paniculatum), when compared to fallows with natural regeneration. In a floodplain ecosystem in central Amazonia, Kreibich et al. [50] found that the flux of mineral N in the soil at 0–20 cm was 3-fold greater under native legume species than under nonlegumes, indicating the importance of legumes in that ecosystem.

A continent-wide study in Africa found that green manuring increases maize production, with important benefits for food security [51, 52]. The great variety of legume species with potential for use as green manures offers a number of possibilities for intercropping. Quick-growing herbaceous species include Cajanus cajan, Crotalaria juncea, Stizolobium aterrimum, and Calopogonium mucunoides, which decompose rapidly and improve soil within 2-3 months [53–55]. In contrast, tree legumes with slower growth, such as Gliricidia sepiumand Leucaena leucocephala, are important for supplying organic matter with different characteristics, such as greater C/N ratios, polyphenol, or lignin contents. While organic matter with a low C/N ratio will decompose more quickly and rapidly releasing nutrients for use by crop plants, in certain situations a slower decomposition rate may be desirable, in order to maintain soil cover and control growth of weeds [56, 57].

However, the diversification of species is important to also diversify the characteristics of the biomass that are generated and incorporated in the soil, such that there is better mix in terms of nutrient contents. This may require planting or managing other species, besides legumes. In the study previously cited by Mochiutti and Queiroz [49] it was found that litter produced in an area with natural secondary forest regeneration had a higher content of Ca than litter from the area planted with the legume taxi-branco. A study by Gonçalvez et al. [58], in the Atlantic Rainforest of Brazil, found marked differences in levels of macro- and micronutrients in litter produced by native tree species, reinforcing the need to consider a mix of species in agroforestry systems.

6. TREES IN AMAZONIAN AGRICULTURE

Over millennia, traditional populations have diversified their agricultural systems with trees, and intentionally or not, have benefitted from their products and other services that result from this management. In Amazonia, many indigenous societies and their multiracial descendants, the caboclos or ribereños, practice forms of agriculture involving trees, whether in homegardens around their dwellings or more extensive systems involving intercropping in swidden fields [59–61]. These practices consist of a set of techniques developed by traditional populations over the years or centuries, based on observation and experimentation.

According to Clement [62], when Europeans first arrived in the Americas, Amazonian communities had already domesticated more than 130 species, of which around 80 are fruit bearing, such as avocado (Persea americana), abiu (Pouteria caimito), biribá (Rollinia mucosa), cashew (Anacardium occidentale), genipap (Genipa americana), papaya (Carica papaya), passion fruit (Passiflora edulis), and murici (Byrsonima crassifolia), among others. It is believed that the domestication of plants and other management practices of tree species originally occurred in areas surrounding dwellings, where discarded seeds of wild species may have germinated, to be cultivated by observant protofarmers [63, 64]. In the locations around dwelling the deposition of residues resulting from domestic activities is commonplace (husks, bones, ashes, etc.) and has led to the creation of some of the most productive soils in Amazonia, the "black earths" or "terra preta de índio" [56, 65–67].

Many of the practices that created such soils are still in use today, as can be observed in the modern-day homegardens of indigenous and caboclo farmers, which have a diversity of tree species, frequent additions of organic matter as well as occasional low-intensity burning of residues [64, 68–73] (Figure 4). In indigenous communities of the Lavrado (savanas) region of Roraima, northern Brazil, homegardens have a great variety of plants, mostly fruit bearing and are important to food security [74]. In a study of soil nutrients under a chronosequence of these homegardens in Araçá Indigenous Land, Roraima, Pinho et al. [75] found a progressive increase in levels of P, K, Ca, Mg, organic matter, and pH, as well as a decrease in Al (Table 1). In these otherwise infertile savanna soils [76], cultivation of fruit trees around dwellings is made possible by this gradual increment in nutrients, a result of nutrient imports from surrounding ecosystems, deposited in the form of residues and possibly the internal cycling by trees that also contribute to the system by means of their litter.

Table 1. Levels of nutrients, organic matter, and acidity, by depth, in homegardens and savanna soils in Araçá Indigenous Land, Roraima, and Brazil. Homegarden values followed by * or ** are statistically different from savannas, with and , respectively, by Mann-Whitney test (-test).

	Savanna (n = 15)	New homegardens (n = 5) (0–10 years-old)	Estabilished homegardens (n = 5) (15–35 years-old)	Old homegardens (n = 5) (40+ years old)
P (mg Kg^{-1})				
0–10 cm	2,1	14,3**	26,5**	43,7**
10–20 cm	1,1	8,5*	12,2**	37,9**
20–30 cm	0,6	7,5*	7,5**	31,2**
K (cmolc Kg^{-1})				
0–10 cm	0,07	0,08	0,11*	0,18**
10–20 cm	0,06	0,08	0,10*	0,15**
20–30 cm	0,06	0,07	0,10*	0,13**
Ca (cmolc Kg^{-1})				
0–10 cm	0,32	0,66*	1,48*	1,93**
10–20 cm	0,22	0,24	1,02*	1,41**
20–30 cm	0,17	0,18	0,63*	1,03**
Mg (cmolc Kg^{-1})				
0–10 cm	0,16	0,18	0,30	0,48**
10–20 cm	0,09	0,10	0,24**	0,33**
20–30 cm	0,06	0,07	0,18*	0,28**
Fe (mg Kg^{-1})				
0–10 cm	25,2	21,4	39,8	135,6*
10–20 cm	22,7	17,8	37,4	104,4*
20–30 cm	20,3	12,6	26,5	81,0*

Zn ($mg\ Kg^{-1}$)				
0–10 cm	1,9	3,1	12,9**	13,9**
10–20 cm	2,1	1,5	8,11**	11,6**
20–30 cm	2,6	1,4	8,63**	11,4*
Mn ($mg\ Kg^{-1}$)				
0–10 cm	22,4	32,4	29,91	40,6
10–20 cm	16,7	23,9	18,02	24,7
20–30 cm	15,3	23,4	14,93	18,5
Cu ($mg\ Kg^{-1}$)				
0–10 cm	0,86	0,74	1,70	1,17
10–20 cm	0,87	0,67	1,37	1,17
20–30 cm	0,84	0,62	1,28	1,13
Organic matter ($g\ Kg^{-1}$)				
0–10 cm	12,9	12,2	20,6	27,6**
10–20 cm	10,0	7,9	14,6	19,5*
20–30 cm	8,3	6,9	10,4	14,3*
Al ($mg\ Kg^{-1}$)				
0–10 cm	0,13	0,07	0,08	0,05*
10–20 cm	0,17	0,12	0,09	0,08
20–30 cm	0,19	0,11	0,11	0,10
pH (H_2O)				
0–10 cm	5,34	5,39	5,48	5,45
10–20 cm	5,16	5,28	5,39	5,49
20–30 cm	5,13	5,29	5,30	5,32

Figure 4. (a) Delimitation between the swept and weeded bare earth "yard" that surrounds the dwelling (left) and the outside area (periphery) where the residues are directed to (right). (b) Accumulation of organic domestic residues in the periphery of a homegarden, where frequent small burns of these residues may occur, together with other organic materials derived from sweeping and weeding the "yard"; (c) outside view of an indigenous homegarden in the "Lavrado" (savanna) of Roraima, Brazil (Araçá Indigenous Land, Brazil).

While dwellings and associated homegardens represent relatively stable features on the landscape, slash-and-burn or swidden agriculture fields shift around a broader area, because for every field under cultivation, a greater area must be kept recovering under a fallow of secondary vegetation. Under low population density, this traditional system is adequate for crop production, with long fallows between cropping cycles guaranteeing the recovery of soil physical and chemical qualities. However, with population growth and pressure on land, more areas of forest are being converted to agriculture, with a concurrent reduction in fallow cycles, such that the sustainability of this system is threatened by reduced soil fertility [14]. In the savanna region of Roraima, intensive agricultural use of the scattered forest patches or islands, with shorter fallow cycles, has led to a reduction in soil fertility and a greater proportion of the area under secondary forest. A study by Freitas [77] showed that in this situation fallows, can contribute to an increase in soil P but not to an increase in K, suggesting that these two nutrients are limiting factors for agricultural production in the forest islands. On the other hand, in a cambisol in the Upper Solimões region, Soares [78] observed that five years of fallow were the minimum time necessary to achieve levels of exchangeable Ca and Mg equivalent to that found in twenty year-old fallows.

Modern agricultural technologies offer the possibility of continuous crop production in the same area, without the need for fallows, through the use of machinery and chemical inputs. However, this model of agricultural production can lead to soil erosion as well to the reduction in biological processes that are important in tropical soils, along with a dependence on external inputs [9]. Based on the soil improvement that naturally occurs under secondary forests, lessons can be obtained as to the possibility of including trees as components of a more sustainable agriculture for Amazonia [79]. Choices as to which tree species to include can consider those that offer useful products or that speed-up processes of soil improvement, such as legumes that provide green manures.

A recently released study classifying land use in areas deforested up to 2007 in the Brazilian Amazon indicates that 62.2% of the area deforested has been occupied by pastures. In productive terms, however, 25% of the area under pasture can be considered as degraded or "weedy" [80]. Such areas left behind the advancing agricultural frontier are now the prime concern for initiatives of landscape recovery that seek to restore ecosystem services, such as riparian buffer strips to protect hydrological resources and provide connectivity for other landscape functions such as biodiversity. Current Brazilian environmental legislation requires that rural properties in Amazonia maintain 80% of their area with forest cover, a figure that is reduced to 50% in regions that have established a zoning system. As a result, many properties must recuperate areas that were deforested if they wish to access government programs such as agricultural credit. While ecological restoration can be quite expensive on its own, recuperation of areas of degraded pasture with agroforestry systems for production of food, commodities, and timber products may be a viable alternative. Choice of the proper tree species is critical to the success of any such initiatives, not just from an economic point of view, but also from the biological point of view in terms of rejuvenating soil properties.

Research with native legumes in Amazonia has shown that more than 60% are nodulating species [81, 82] and thus have a great potential for use in agroforestry. Other tree species may be chosen based on either ecological or production functions. However, planning of successful agroforestry systems must alsoconsider cash flows and choose appropriate species accordingly.

Although the number of practical examples is limited, another potential role of trees in Amazonian production systems is as components of silvopastoral systems. Andrade et al. [83] indicate that at least one legume species, Stryphnodendron guianensis, has positive effects on soil fertility and forage production. However, this is due in part to this tree's crown architecture, which does not create an excess of shade.

7. CONCLUSIONS

Trees in agroeco systems can be present in an infinite number of arrangements and species combinations that depend mostly on farmers' objectives as well as the environmental characteristics of the region. In general, it is safe to say that a greater diversity of species is more favorable, as it results in a more complete occupation of space above and below the soil, and the variation in the characteristics of the litter produced can maintain a greater level of soil biodiversity, with positive effects on fertility [10, 79].

The intentional use of trees in farming systems occurs in many parts of the globe; however, many of their benefits, that go much beyond soil improvement, are still difficult to visualize or quantify in economic terms or only appear in the long term [47, 84], such that many farmers are hesitant to invest in trees. A better understanding of the positive effects of trees on soils, and an economic analysis of what this represents in terms of nutrients and other benefits, is an important step towards increasing the use of trees on farms.

Soil improvement under trees and agroforestry systems is in great part related to increases in organic matter, whether in the form of surface litter or soil carbon. Therefore, besides their role in above-ground carbon sequestration, agroforestry systems also have a great potential to increase carbon stocks in the soil and certainly merit consideration in mechanisms that propose payments for mitigation of greenhouse gas emissions to reduce climate change.

REFERENCES

1. R. J. Zomer, A. Trabucco, R. Coe, and F. Place, Trees on Farm: Analysis of Global Extent and Geographical Patterns of Agroforestry, ICRAF Working Paper no. 89, World Agroforestry Centre, Nairobi, Kenya, 2009.

2. World Agroforestry Centre, 2008,

3. R. Porro, Ed., Alternativa agroflorestal na Amazônia em transformação, Embrapa Informação Tecnológica, Brasília, Brazil, 2009.

4. P. K. R. Nair, Agroforestry Systems in the Tropics, Kluwer, Norwell, Mass, USA, 1989.

5. R. E. Ricklefs, A Economia da Natureza, Guanabara Koogan, Rio de Janeiro, Brazil, 3rd edition, 1996.

6. A. Primavesi, "A fertilidade do solo," Agroecologia Hoje, vol. 8, article 5, 2001.

7. F. J. Luizao, "Litter production and mineral element input to the forest floor in a Central Amazonian forest," GeoJournal, vol. 19, no. 4, pp. 407–417, 1989.

8. P. M. Fearnside and R. I. Barbosa, "Soil carbon changes from conversion of forest to pasture in Brazilian Amazonia," Forest Ecology and Management, vol. 108, no. 1-2, pp. 147–166, 1998.

9. A. Primavesi, Manejo ecológico do solo: a agricultura em regiões tropicais, Nobel, São Paulo, Brazil, 2002.

10. F. J. Luizão and R. C. C. Luizão, "Matéria orgânica do solo em Roraima," in Homem, Ambiente e Ecologia no Estado de Roraima, R. I. Barbosa, E. J. G. Ferreira, and E. G. Castellón, Eds., pp. 363–376, INPA, Manaus, Brazil, 1997.

11. S. J. F. Ferreira, F. J. Luizão, S. Á. F. Miranda, M. D. S. R. Da Silva, and A. R. T. Vital, "Nutrients in soil solution in an upland forest submitted to selective logging in central Amazonia," Acta Amazonica, vol. 36, no. 1, pp. 59–68, 2006.

12. P. J. Kleinman, D. Pimentel, and R. B. Bryant, "The ecological sustainability of slash-and-burn agriculture," Agriculture, Ecosystems and Environment, vol. 52, no. 2-3, pp. 235–249, 1995.

13. P. Santilli, "Ocupação territorial Macuxi: aspectos históricos e políticos," in Homem, Ambiente e Ecologia no Estado de Roraima, R. I. Barbosa, E. J. G. Ferreira, and E. G. Castellón, Eds., pp. 49–64, INPA, Manaus, Brazil, 1997.

14. M. Mazoyer and L. Roudart, História das agriculturas no mundo: do neolítico à crise contemporânea, UNESP, São Paulo and NEAD, Brasília, Brazil, 2010.

15. P. J. Zinke, "The pattern of influence of individual forest trees on soil properties," Ecology, vol. 43, pp. 130–133, 1962.

16. A. J. Belsky, R. G. Amundson, J. M. Duxbury, S. J. Riha, A. R. Ali, and S. M. Mwonga, "The effects of trees on their physical, chemical, and biological environments in a semi-arid savanna in Kenya," Journal of Applied Ecology, vol. 26, no. 3, pp. 1005–1024, 1989.

17. I. C. Burke, W. K. Lauenroth, M. A. Vinton et al., "Plant-soil interactions in temperate grasslands,"Biogeochemistry, vol. 42, no. 1-2, pp. 121–143, 1998.

18. W. H. Schlesinger, J. A. Raikks, A. E. Hartley, and A. F. Cross, "On the spatial pattern of soil nutrients in desert ecosystems," Ecology, vol. 77, no. 2, pp. 364–374, 1996.

19. W. E. Frost and S. B. Edinger, "Effects of tree canopies on soil characteristics of annual rangeland,"Journal of Range Management, vol. 44, no. 3, pp. 286–288, 1991.

20. A. L. Ulery, R. C. Graham, O. A. Chadwick, and H. B. Wood, "Decade-scale changes of soil carbon, nitrogen and exchangeable cations under chaparral and pine," Geoderma, vol. 65, no. 1-2, pp. 121–134, 1995.

21. A. C. Finzi, C. D. Canham, and N. Van Breemen, "Canopy tree-soil interactions within temperate forests: species effects on pH and cations," Ecological Applications, vol. 8, no. 2, pp. 447–454, 1998.

22. F. A. Dijkstra, "Calcium mineralization in the forest floor and surface soil beneath different tree species in the northeastern US," Forest Ecology and Management, vol. 175, no. 1–3, pp. 185–194, 2003.

23. P. B. Reich, J. Oleksyn, J. Modrzynski et al., "Linking litter calcium, earthworms and soil properties: a common garden test with 14 tree species," Ecology Letters, vol. 8, no. 8, pp. 811–818, 2005.

24. R. J. Scholes and S. R. Archer, "Tree-grass interactions in Savannas," Annual Review of Ecology and Systematics, vol. 28, pp. 517–544, 1997.

25. E. G. Jobbágy and R. B. Jackson, "The distribution of soil nutrients with depth: global patterns and the imprint of plants," Biogeochemistry, vol. 53, no. 1, pp. 51–77, 2001.

26. M. Arianoutsou, "Atmospheric deposition of nutrients in a coastal maquis ecosystem of northeastern Greece," International Journal of Biometeorology, vol. 33, no. 2, pp. 124–130, 1989.

27. C. Bertin, X. Yang, and L. A. Weston, "The role of root exudates and allelochemicals in the rhizosphere,"Plant and Soil, vol. 256, no. 1, pp. 67–83, 2003.

28. P. K. R. Nair, V. D. Nair, B. Mohan Kumar, and J. M. Showalter, "Carbon sequestration in agroforestry systems," Advances in Agronomy, vol. 108, pp. 237–307, 2010.

29. P. K. R. Nair, B. M. Kumar, and V. D. Nair, "Agroforestry as a strategy for carbon sequestration," Journal of Plant Nutrition and Soil Science, vol. 172, no. 1, pp. 10–23, 2009.

30. A. Takimoto, P. K. R. Nair, and V. D. Nair, "Carbon stock and sequestration potential of traditional and improved agroforestry systems in the West African Sahel," Agriculture, Ecosystems and Environment, vol. 125, no. 1–4, pp. 159–166, 2008.

31. S. K. Saha, P. K. R. Nair, V. D. Nair, and B. M. Kumar, "Carbon storage in relation to soil size-fractions under tropical tree-based land-use systems," Plant and Soil, vol. 328, no. 1, pp. 433–446, 2010.

32. D. S. Howlett, M. R. Mosquera-Losada, P. K. R. Nair, V. D. Nair, and A. Rigueiro-Rodrigues, "Soil carbon storage in silvopastoral systems and a treeless pasture in northwestern Spain," Journal of Environmental Quality, vol. 40, no. 3, pp. 825–832, 2011.

33. R. G. Tonucci, P. K. R. Nair, V. D. Nair, R. Garcia, and F. S. Bernardino, "Soil carbon storage in silvopasture and related land-use systems in the Brazilian Cerrado," Journal of Environmental Quality, vol. 40, no. 3, pp. 833–841, 2011.

34. R. Derpsch, "No-tillage and conservation agriculture: a progress report," in No-Till Farming Systems, T. Goddard, M. Zoebisch, Y. Gan, W. Ellis, A. Watson, and S. Sombatpanit, Eds., pp. 7–39, World Association of Soil and Water Conservation. Special Publication No. 3, 2008.

35. P. Lavelle and B. Pashanasi, "Soil macrofauna and land management in Peruvian Amazonia (Yurimaguas, Loreto)," Pedobiologia, vol. 33, no. 5, pp. 283–292, 1989.

36. P. Lavelle, A. Chauvel, and C. Fragoso, "Faunal activity in acid soils," in Plant Soil Interactions at Low pH, R. A. Date, Ed., pp. 201–211, Springer, Berlin, Germany, 1995.

37. E. Barros, J. Mathieu, S. Tapia-Coral, A. R. L. Nascimento, and P. Lavelle, "Comunidades da macrofauna do solo da Amazônia Brasileira," in Biodiversidade do Solo em Ecossistemas Brasileiros, F. M. S. Moreira, J. O. Siqueira, and L. Brussaard, Eds., pp. 171–191, Editora UFLA, Minas Gerais, Brazil, 2008.

38. S. C. Tapia-Coral, F. J. Luizão, and E. V. Wandelli, "Macrofauna da liteira em sistemas agroflorestais sobre pastagens abandonadas na Amazônia Central," Acta Amazônica, vol. 29, no. 3, pp. 447–495, 1999.

39. P. L. Leal, S. L. Stürmer, and J. O. Siqueira, "Occurrence and diversity of arbuscular mycorrhizal fungi in trap cultures from soils under different land use systems in the amazon, Brazil," Brazilian Journal of Microbiology, vol. 40, no. 1, pp. 111–121, 2009.

40. E. D. C. Jesus, T. L. Marsh, J. M. Tiedje, and F. M. D. S. Moreira, "Changes in land use alter the structure of bacterial communities in Western Amazon soils," ISME Journal, vol. 3, no. 9, pp. 1004–1011, 2009.

41. P. Lavelle, E. Barros, E. Blanchart et al., "SOM management in the tropics: why feeding the soil macrofauna?" Nutrient Cycling in Agroecosystems, vol. 61, no. 1-2, pp. 53–61, 2001.

42. V. V. Souza, R. C. Pinho, and I. Jucksch, "Produção e composição de serrapilheira em um sistema agroflorestal com café (Coffea arabica) em Viçosa-MG," in Anais do V Congresso Brasileiro de Sistemas Agroflorestais, Curitiba, Brazil, 2004.

43. G. P. Araujo and L. S. Collier, "Parâmetros de fertilidade dos solos em sistemas agroflorestais em Esperantina-TO," in Anais do VI Congresso Brasileiro de Sistemas Agroflorestais, Campos, Brazil, 2006.

44. C. Jaramillo-Botero, R. H. S. Santos, Junior P. Marco, T. M. Pontes, P. Fardin, and F. Sarmento, "Efeito da distribuição espacial das árvores sobre a produção de cafeeiros em sistema agroflorestal na zona da mata de Minas Gerais, Brasil," in Anais do VI Congresso Brasileiro de Sistemas Agroflorestais, Campos, Brazil, 2006.

45. R. C. Pinho, V. V. Souza, and I. Jucksch, "Contribuição de espécies arbóreas na ciclagem de nutrientes via precipitação em um sistema agroflorestal com café (Coffea arabica) no município de Viçosa, MG," inAnais do IV Congresso Brasileiro de Sistemas Agroflorestais, Ilhéus, Brazil, 2002.

46. J. Jesus, M. S. Bernardes, C. A. Righi, A. M. P. Lunz, J. L. Favarin, and F. T. Camargo, "Avaliação da fertilidade do solo e teor foliar de K do cafeeiro (Coffea arabica L.) em sistema agroflorestal em aléia de seringueira (Hevea brasiliensis Muell. Arg.) e em monocultivo," in Proceedings of the Anais do VI Congresso Brasileiro de Sistemas Agroflorestais, Campos, Brazil, 2006.

47. G. Schroth, J. Lehmann, M. R. L. Rodrigues, E. Barros, and J. L. V. Macêdo, "Plant-soil interactions in multistrata agroforestry in the humid tropics," Agroforestry Systems, vol. 53, no. 2, pp. 85–102, 2001.

48. P. M. Vitousek, K. Cassman, C. Cleveland et al., "Towards an ecological understanding of biological nitrogen fixation," Biogeochemistry, vol. 57-58, pp. 1–45, 2002.

49. S. Mochiutti and J. A. L. Queiroz, "Aporte de nutrientes ao solo via serrapilheira em pousios florestais com taxi-branco e capoeira no Amapá," in Anais do VI Congresso Brasileiro de Sistemas Agroflorestais, Campos, Brazil, 2006.

50. H. Kreibich, J. Lehmann, G. Scheufele, and J. Kern, "Nitrogen availability and leaching during the terrestrial phase in a várzea forest of the Central Amazon floodplain," Biology and Fertility of Soils, vol. 39, no. 1, pp. 62–64, 2003.

51. C. Pye-Smith, Farming trees, banishing hunger. How an agroforestry programme is helping smallholders in Malawi to grow more food and improve their livelihoods, World Agroforestry Centre, Nairobi, Kenya, 2008.

52. G. Sileshi, F. K. Akinnifesi, O. C. Ajayi, and F. Place, Evidence for impact of green fertilizers on maize production in sub-Saharan Africa: a meta-analysis, ICRAF Occasional Paper No. 10, World Agroforestry Centre, Nairobi, Kenya, 2009.

53. F. A. De Alcântara, A. E. Furtini Neto, M. B. De Paula, H. A. De Mesquita, and J. A. Muniz, "Green manuring in the recovery of degraded oxisoil fertility," Pesquisa Agropecuaria Brasileira, vol. 35, no. 2, pp. 277–288, 2000.

54. L. R. Queiroz, F. C. Coelho, D. G. Barroso, and V. A. V. Queiroz, "Evaluation of phytomass productivity and N, P and K accumulation of shrub legumes in alley cropping system in Campos dos Goytacazes (RJ)," Revista Arvore, vol. 31, no. 3, pp. 383–390, 2007.

55. E. A. Delarmelinda, F. A. R. Sampaio, J. R. M. Dias, L. B. Tavella, and J. S. de Silva, "Green manure and changes on chemical characteristics of a soil in the Ji-Paraná-RO region," Acta Amazonica, vol. 40, no. 3, pp. 625–628, 2010.

56. E. G. Neves, J. B. Petersen, R. N. Bartone, and C. A. Silva, "Historical and socio-cultural origins of Amazonian dark earths," in Amazonian Dark

Earths: Origin, Properties, Management, J. Lehmann, D. Kern, and B. Glaser, Eds., pp. 29–50, Kluwer Academic Publishers, 2003.

57. C. L. Bergo, E. P. Pacheco, H. A. De Mendonça, and J. T. D. S. Marinho, "Evaluation of legume species for coffee plants formation in the segment of family farms in Acre," Acta Amazonica, vol. 36, no. 1, pp. 19–24, 2006.

58. L. V. C. Gonçalves, F. G. B. Barreto, L. C. G. Gomes, E. M. G. Duarte, and I. M. C. Cardoso, "Contribuições de árvores nativas da Mata Atlântica para a ciclagem de nutrientes em sistemas agroflorestais," in Anais do Simpósio Mineiro de Ciência do Solo, Viçosa, Brazil, 2010.

59. D. A. Posey, "Manejo da floresta secundária, capoeiras, campos e cerrados (Kayapó)," in Suma Etnológica Brasileira, pp. 173–185, FINEP, 1987.

60. B. G. Ribeiro, "Classificação dos solos e horticultura Desâna," in 1st International Congress of Ethnobiology, vol. 2, pp. 27–49, Belém, Brazil, 1990.

61. R. P. Miller and P. K. R. Nair, "Indigenous agroforestry systems in Amazonia: from prehistory to today,"Agroforestry Systems, vol. 66, no. 2, pp. 151–164, 2006.

62. C. R. Clement, "1492 and the loss of amazonian crop genetic resources. I. The relation between domestication and human population decline," Economic Botany, vol. 53, no. 2, pp. 188–202, 1999.

63. D. W. Lathrap, "Our father the cayman, our mother the gourd: Spinden revisited, or a unitary model for the emergence of agriculture in the New World," in Origins of Agriculture, pp. 713–751, Mouton, The Hague, The Netherlands, 1977.

64. R. P. Miller, J. W. Penn, and J. Leeuwen, "Amazonian homegardens: their ethnohistory and potential contribution to agroforestry development," in Tropical Homegardens: A Time-Tested Example of Sustainable Agroforestry, B. M. Kumar and P. K. R. Nair, Eds., pp. 43–60, Springer, 2006.

65. C. Erickson, "Historical ecology and future explorations," in Amazonian Dark Earths: Origin, Properties, Management, J. Lehmann, D. Kern, and

B. Glaser, Eds., pp. 455–500, Kluwer Academic Publishers, Dodrecht, The Netherlands, 2003.

66. B. Glaser, W. Zech, and W. I. Woods, "History, current knowledge and future perspectives of geoecological research concerning the origin of Amazonian anthropogenic dark earths (Terra Preta)," inExploration in Amazonian Dark Earths, B. Glaser and W. I. Woods, Eds., pp. 9–17, Springer, Berlin, Germany, 2004.

67. A. B. Junqueira, G. H. Shepard, and C. R. Clement, "Secondary Forests on Anthropogenic Soils of the Middle Madeira River: Valuation, Local Knowledge, and Landscape Domestication in Brazilian Amazonia," Economic Botany, vol. 65, no. 1, pp. 85–99, 2011.

68. B. M. Kumar, S. J. George, and S. Chinnamani, "Diversity, structure and standing stock of wood in the homegardens of Kerala in peninsular India," Agroforestry Systems, vol. 25, no. 3, pp. 243–262, 1994.

69. R. M. B. Lima, Descrição, Composição e Manejo dos Cultivos Mistos de Quintal naVárzea da "Costa do Caldeirão", M.S. dissertation, INPA/UFAM, Manaus, Brazil, 1994.

70. E. Brocki, Sistemas agroflorestais de cultivo e pousio: etnoconhecimento de agricultores familiares do lago do Paru (Manacapuru, AM), Tese de doutorado, INPA/UA, Manaus, Brazil, 2001.

71. W. M. Denevan, "Semi-intensive pre-european cultivation and the origins of anthropogenic dark earths in Amazonia," in Exploration in Amazonian Dark Earths, B. Glaser and W. I. Woods, Eds., pp. 135–143, Springer, Berlin, Germany, 2004.

72. A. Peyre, A. Guidal, K. F. Wiersum, and F. Bongers, "Dynamics of homegarden structure and function in Kerala, India," Agroforestry Systems, vol. 66, no. 2, pp. 101–115, 2006.

73. R. C. Pinho, Quintais agroflorestais indígenas em área de savana (Lavrado) na Terra Indígena Araçá, Roraima, dissertation, INPA/UFAM, Manaus, Brazil, 2008.

74. R. C. Pinho, L. D. Magalhães, R. P. Miller, K. Uguen, and S. S. Alfaia, "Description of homegardens in Araçá Indigenous Land, in the Lavrado

(savannas) of Roraima, Brazil," in 2nd World Congress of Agroforestry, Nairobi, Kenya, 2009.

75. R. C. Pinho, S. S. Alfaia, R. P. Miller et al., "Islands of fertility: soil improvement under indigenous homegardens in the savannas of Roraima, Brazil," Agroforestry Systems, vol. 81, no. 3, pp. 235–247, 2011.

76. J. F. Vale Júnior and M. I. L. Souza, "Caracterização e distribuição dos solos das savanas de Roraima," inSavanas de Roraima: Etnoecologia, Biodiversidade e Potencialidades Agrossilvipastoris, R. I. Barbosa, H. A. M. Xaud, and J. M. Costa e Sousa, Eds., pp. 79–92, FEMACT, Boa Vista, Brazil, 2005.

77. V. M. B. Freitas, Dinâmica do nitrogênio em capoeiras e florestas em savanas de Roraima, M.S. dissertation, Instituto Nacional de Pesquisas da Amazônia, Manaus, Brazil, 2008.

78. A. E. S. Soares, Nutrientes e Carbono no Solo em Áreas com diferentes Sistemas de Uso na Região do Alto Solimões (Benjamin Constant AM), M.S. dissertation, Instituto Nacional de Pesquisas da Amazônia, 2006.

79. J. L. Vivan, Agricultura e Florestas: Princípios de uma Integração Vital, Agropecuária, Guaíba, Brazil, 1998.

80. Empresa Brasileira de Pesquisa Agropecuária—Embrapa and Instituto Nacional de Pesquisas Espaciais—INPE, Levantamento de informações de uso e cobertura da terra na Amazônia, Sumário Executivo, Brasilia, Brazil, 2011.

81. F. M. M. Magalhães, L. M. S. Magalhães, L. A. Oliveira, and J. Dobereiner, "Ocorrência de nodulação em leguminosas florestais de terra firme nativas da região de Manaus—AM," Acta Amazonica, vol. 12, no. 3, pp. 509–514, 1982.

82. L. A. G. Souza, M. F. Silva, and F. W. Moreira, "Capacidade de nodulação de cem leguminosas da Amazônia," Acta Amazonica, vol. 24, no. 1-2, pp. 9–18, 1994.

83. C. M. S. De Andrade, J. F. Valentim, and J. D. C. Carneiro, "Baginha trees (Stryphnodendron guianense(Aubl.) Benth.) in cultivated pasture

ecosystems in the Western Amazon," Revista Brasileira de Zootecnia, vol. 31, no. 2, pp. 574–582, 2002.

84. A. M. N. Izac, "Economic aspects of soil fertility management and agroforestry practices," in Trees, Crops and Soil Fertility, G. Schroth and F. L. Sinclair, Eds., pp. 13–37, CAB International, 2003.

Chapter 5

Integration of Commercial Microbiological Products into Soil Fertility Practices as a Potential Option for Acclimatization and Growth of TC Banana in Kenya

Agnes Mumo Kavoo-Mwangi[1,2,3], Esther M. Kahangi[2], Elijah Ateka[2], Justus Onguso[2], Joyce M. Jefwa[3]

[1]Crop Science Department, School of Agriculture and Biotechnology, Karatina University, Karatina, Kenya

[2]Department of Horticulture, School of Agriculture, Jomo Kenyatta University of Agriculture and Technology, Nairobi, Kenya

[3]Tropical Soil Biology and Fertility Institute of the International Centre for Tropical Agriculture (TSBF-CIAT), Nairobi, Kenya

ABSTRACT

Tissue culture (TC) banana plantlets at the in vitro stage are delicate and devoid of microbes and nutrients that are essential for establishment and subsequent growth. Some microbes are known for function best under certain soil threshold levels of macro and micronutrients and have been associated with growth and performance of TC banana. A green house and field study was conducted to evaluate the effect of combining two commercial biological products [Rhizatech and ECO-T (mycorrhiza and Trichoderma based products, respectively)] with various sources of nitrogen and phosphorous including Mavuno, Minjingu phosphate rock, Calcium Ammonium Nitrate (CAN), manure and diammonium phosphate (DAP) on growth and performance of TC banana in Vertisol and Rhodic Ferralsol soil conditions. Tissue culture plants were initially inoculated with Rhizatech and ECO-T at the acclimatization stage and subsequently at the beginning of the potting stage and field establishment. Addition of nutrient sources was also done at the same stages of plant growth by mixing with the soil substrates prior to planting. The performance of plants was significantly (at $p \leq 0.05$) affected by the combinations of nutrient sources depending on the soil type and stage of plant development. The growth of plants in the Vertisol increased with Trichoderma combined with either organic manure, DAP or combined with a macro and micro nutrient source (Mavuno) as compared to the sole application of Trichoderma. Performance of plants treated with combination of mycorrhiza and either Mavuno and minjigu rock phosphate was consistently higher in the Rhodic Ferralsol than either mycorrhiza alone or fertilizer alone. This indicates that TC plants could highly benefit from combined application of microbiological products and inorganic and organic fertilizers. However, a prior knowledge of the product's microbial formulation and prevailing soil conditions is essential for optimizing the potential benefits of integrating microbe-based product with inorganic and organic fertilizers.

KEYWORDS

Microbiological Products, Soil Fertility Practices, Integration, Tissue Culture Banana, Growth and Performance

1. INTRODUCTION

Tissue culture banana plantlets offer an excellent means for providing pestand disease-free planting material to farmers [1] [2]. However, tissue culture (TC) plants are fragile, devoid of food reserves and therefore more prone to shock, which may lead to loss of plantlets at establishment. With decline in soil fertility and increase in soilborne pests and diseases, TC plants are likely to succumb to disease and roots may not establish well in low fertility soil environments [3].

Banana requires large amounts of nitrogen and potassium followed by phosphorus, calcium and magnesium to maintain high yields [4] [5]. To fulfill the plant demand for nutritional attributes, it is essential to apply those elements in the soil which mostly come from inorganic chemical sources. The increased use of chemical fertilizer is undesirable because its production is an energetically costly process and considerable pollution is caused through both the production and use of mineral N-fertilizers. This is exacerbated by the relatively low efficiency of their uptake by the plants due to non-extensive root system and may also delete soil organic matter in the long term [6] -[8]. Inoculant biofertilizers are more environmentally sound and their introduction in agricultural production systems could be one of the means to mitigate the onset of global warming as well as the reduction in fertilizer input costs, prevent depletion of organic matter and increase crop yields[9] -[11].

Inoculation of tissue culture banana with biofertilizers such as mycorrhiza (AMF) and Trichoderma has been reported to stimulate root growth, nutrient uptake, enhance plant establishment and increase growth, yield and protection against disease and pest infestation [12] -[16]. The growth vigor acquired by biologically inoculated TC banana plants under nursery conditions is expected to give the plants an advantage during field establishment especially in nutritionally and disease challenged soils. However, the functionality of some of the rhizospheric microbes used for constituting biofertilizers such as mycorrhiza (AMF) and Trichoderma is greatly influenced by the prevailing soil conditions. Nitrogen (N), phosphorous (P) and other nutrients such as zinc, copper and sulphur levels in the soil affect the functioning of these microorganisms [17] -[21]. Mycorrhizal symbiosis benefits, for example, have been suggested to be optimal at P levels of 50 mg·kg^{-1} [21].

Organic and inorganic fertilizers are used primarily to increase nutrient availability and the type or amount of fertilizer added to soil could directly affect the function performed by the various microbial groups in the soil [22]. Mineral nutrition is essential for growth, sporulation and stimulation of fungal secondary metabolism [18] and combining of mineral and bio fertilizers could greatly benefit both the added microbes and inoculated plant as well as improve soil quality [11][23] -[25].

Previous studies have focused on the effect of long term fertilization on indigenous microbial diversity and efficiency [26] [27] but little is reported on the effect of combining commercial microbiological strains with fertilizers in nutrient poor soils that have little or no history of fertilization.

Although several formulations of mycorrhiza and Trichoderma are available in the market, their efficacy on plant growth is variable depending on the soil nutrient levels. It is therefore essential to add these nutrients in levels that enable optimal functioning of the micro-organisms and consequent promotion of nutrient uptake and plant growth. The integration of microbiological products with fertilizers could address the soil fertility constraint faced by TC banana plantlets, especially under field conditions. This study focused on evaluating the potential of integrating microbiological products with inorganic and organic fertilizers on growth and performance of tissue cultured banana in Vertisol and Rhodic Ferralsol soil conditions under nursery and field conditions.

2. MATERIALS AND METHODS

2.1. Source of Tissue Cultured Materials, Soil Properties and Inoculation Process

Tissue culture banana plantlets cv. Gros Michel, were obtained from Jomo Kenyatta University of Agriculture and Technology (JKUAT) Biotechnology laboratory.

The experimental soils (Vertisol and Rhodic Ferralsol) were sampled from two banana growing regions in Kenya namely, Western Kenya (Bondo) and Coastal Kenya (Kilifi) (Table 1) at a depth of 0 - 20 cm and used for hardening and potting of tissue culture plantlets. Soil nutrient composition (Nitrogen, Phosphorous, Potassium, Carbon, Magnesium, Calcium, Sodium), Cation Exchange Capacity (CEC), pH and soil texture composition (% Clay, % Sand and % Silt) were determined according to standard procedures [28] [29]. The initial soil characteristics are described in Table 2

2.2. Green House Experiment

A 2 × 2 × 3 factorial experiment comprising of 1) two microbiological products a) Rhizatech b) ECO-T; 2) soil substrates a) Vertisol (sampled from Western Kenya-Bondo) b) Rhodic Ferralsol (sampled from coastal Kenya-Kilifi) and; 3) Fertilizers a) Minjingu phosphate rock (MPR) and CAN as sources of phosphorous (P), Mg, Ca and eight more essential micronutrients (Mwaluko, 1996) and nitrogen (N), b) Mavuno as a source of N, P, potassium (K) and micronutrients such as calcium Ca, Mg, sulphur (S) and other essential micronutrients including boron (B), manganese (Mn), zinc (Zn), molybdate (Mo) and copper (Cu) c) manure and diammonium phosphate (DAP) as a source of P, N and carbon (C). Three replicates consisting of 16 plants per replicate were considered per treatment. Control experimental units (without product addition) were also included with three replicates per soil. The experimental units were arranged in a completely randomized design under greenhouse conditions. The microbiological products are described in Table 3 Application of products was done as recom mended by the manufacturers. Application rates per plant corresponded to the recommended rate in the field, converted using the medium planting density for banana (2500 plants·ha^{-1}). The area of pots used for establishing plants was considered when determining the exact amount of inoculum to be added per pot. The mode of application of products is described in Table 3 ECO-T was applied at 1.25 g/L, and Rhizatech at 2 g/plant. Three replicates consisting of 10 plants per replicate were considered per treatment. Fertilizer application rates followed the recommended fertilizer rates and were applied at the hardening, potting and at field establishment by mixing with the soil substrate. The areas of pot used for establishing plants were considered when determining the exact amount of microbiological inoculum to be added per pot.

2.3. Inoculation Process

Inoculation of plantlets with products, soil sterilization (Vertisol and Rhodic Ferralsol soils), acclimatization and maintenance under high humidity conditions was done according to standard procedures [30]. The inoculation strategy applied for initial inoculation of deflasked plantlets at the acclimatization stage and subsequently at the potting (8 weeks after deflasking) and field establishment stages (22 weeks after deflasking) is shown in Figure 1.

Table 1. Description of soils used for hardening and potting of tissue culture banana.

Site of soil collection	Description of soil
Western Kenya (Bondo)	Vertisol: dark montmorillonite with swelling properties, heavy cracking when dry; poorly drained, leaching of weathering minerals limited, high in Ca and Mg, pH. Low hydraulic conductivity; N, P and micronutrient deficient; no P-fixation [33].
Coastal Kenya (Kilifi)	Rhodic Ferralsol: developed on limestone, reddish brown to weak red top soil, weak to moderate texture silty clay texture with low nutrient status, especially low in N and P. Deeply weathered soil, moderately fertile, texture loamy sandy with slightly acid pH [33].

Modified from [30].

Table 2. Initial soil characteristics (0 - 20 cm) of Rhodic Ferralsol and Vertisol, used for establishing tissue culture banana under greenhouse conditions.

Soil property*	Units	Rhodic Ferralsol	Vertisol
pH (H_2O)		6.9	5.87
Olsen P	($mg \cdot P \cdot kg^{-1}$)	7	3
K	($cmolc \cdot kg^{-1}$)	0.6	0.82
Ca	($cmolc \cdot kg^{-1}$)	4	25.99
Mg	($cmolc \cdot kg^{-1}$)	1.7	12.53
Na	($cmolc \cdot kg^{-1}$)	0.1	0.19
ECEC	($cmolc \cdot kg^{-1}$)	7	52.5
N	(%)	0.1	0.25
C	(%)	1	3.67
Clay	(%)	19.7	54.7
Sand	(%)	76.3	20.3
Silt	(%)	4.0	25

Modified from [30].

Table 3. Description of microbe-based commercial products used in the experiments.

Product	Manufacturer	Composition	Asserted non-active ingredients	Dose	Mode of application
ECO-T	Plant Health Products (Pty) Ltd., South Africa Distributed by Lachlan Ltd., Kenya	Spores of *Trichoderma* harzianum strain Rifai KRL AG2 (4.8%)	Kaolin clay (95.2%)	0.25 g per 3 m^2	Dry powder applied to soil
Rhizatech	Dudutech (K) Ltd., Kenya	Spores and mycelial fragments of arbuscular mycorrhizal fungi species: *Glomus mosseae*, *G. etunicatum*, *G. intraradices* and *G. aggregatum*). (50 propagules/cm^3)	Not specified	60 $kg{\cdot}ha^{-1}$	Granules applied to soil

CFU, colony forming units. PHC Colonize AG coded as PHC Colonize in the manuscript (modified from [30]).

2.4. Study Site Description and Experimental Design under Field Conditions

The experiment was carried out at two sites with similar soil conditions as those used for hardening and potting of plants at under green house nursery conditions. Tissue cultured banana plants established under green house conditions were used as planting materials and established in two agro ecological zones (Western Kenya in Bondo and Coastal Kenya in Kilifi). The experimental design followed a multilocational complete block design in four farms per study site. Each farm was considered as a replicate. Diammonium phosphate (DAP) and Minjigu rock phosphate fertilizers, used as sources of phosphorous were applied at the rate of 50 kg per hectare. Calcium Ammonium Nitrate (CAN) used as a source of nitrogen was applied at the rate of 120 kg per hectare. Mavuno fertilizer which was used as a source of macro and micro nutrients was applied at the manufacturers recommended rate of 185 kg per hectare. High quality farm yard manure was considered as a source of organic nitrogen and carbon and was applied at 6 tonnes per hectare. Application rates per plant corresponded to the recommended rate in the field, converted using the medium planting density for banana (2500 plants·ha^{-1}).

2.5. Planting and Crop Management

Medium planting density of 2500 plants per hectare was adopted hence the plants were spaced at 2 m by 2 m and established in 60 × 60 × 60 cm planting pits. Guard row plants were established three meters away from the experimental plants only in edge plots, hence surrounding the entire farm. The inner 2 m × 2 m inner mats were monitored and evaluated up to the on-set of flowering of the mother plant. Fertilizer was mixed with top soil and incorporated around the banana plant in the pit followed by topping with sub-soil before planting. Hand weeding was done when needed avoiding the use of tools that could mutilate superficial banana roots. Desuckering of plants was carried out in order to have a 1-2-3 system (mother, first ratoon, second ratoon), preferably with a circular movement of plants, so that original planting density was maintained. Deleafing (leaves having <50% of functional leaf surface area) and watering during periods of severe drought stress were also part of the management practices carried out.

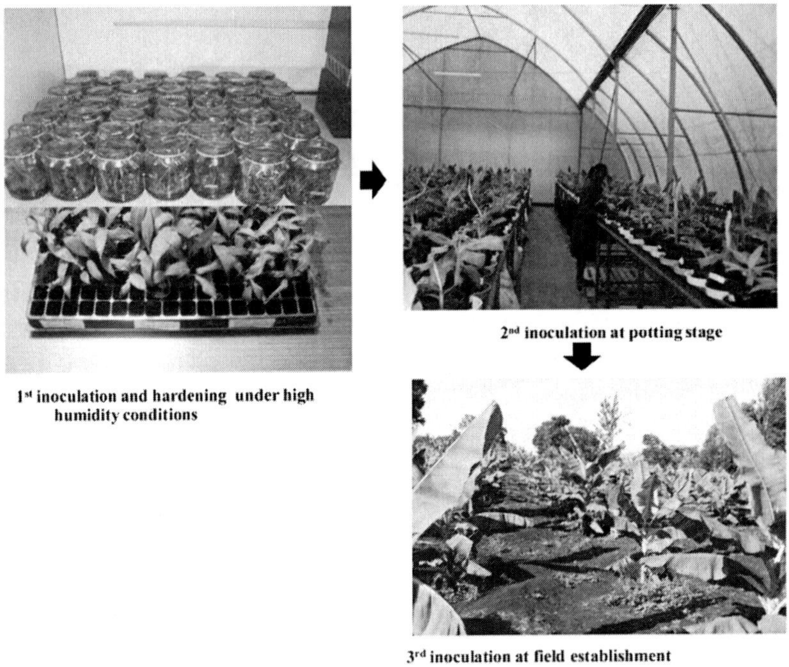

Figure 1. Strategy for inoculating micropropagated banana plantlets.

2.6. Assessment of Growth Parameters

Plant growth parameters such as shoot height (from base), length and width of the second leaf from the sword shoot, girth (at two centimeters from the base) of the stem, and number of functional leaves were recorded every two weeks under greenhouse conditions and every two months under field conditions up to the on-set of flowering stage. Only treatments with a high number of surviving plants under green house conditions were considered for field establishment.

2.7. Data Analysis

To assess the effects of products, soils, fertilizers and the interaction between soils, products and fertilizers, data were subjected to analyses of variance (ANOVA) using the mixed procedure of SAS[31]. Means found to be significant at $p \leq 0.05$ were separated using Tukey's HSD test.

3. RESULTS

3.1. Integration of Microbiological Products with Fertilizers on Growth and Performance of Tissue Cultured Plants

The effect of combining products with various fertilizers was evident in both Vertisol and Rhodic Ferralsol. Highly significant ($p \leq 0.001$) differences between treatments were observed on all plant growth parameters evaluated in both soils (Table 4 and Table 5). In the Rhodic Ferralsol, the highest plant height was observed on plants treated with combined application of Rhizatech and Mavuno with an increase of 136% and 116.5% compared to the non-treated control and sole application of Rhizatech, respectively (Table 4). The highest number of leaves (10), which was statistically different from all other treatments, was observed with the combined application of minjigu rock phosphate and CAN (MRPCAN). Plants treated with combined application of ECO-T and MRPCAN recorded the highest increase in girth of 55% compared to the non-treated control, 44% and 63% compared to the application of MRPCAN alone and ECO-T alone. Leaf length was statistically different between most treatments with the longest leaf length observed on plants treated with combined application of ECO-T and manure and DAP. The combined application of ECO-T and manure and DAP increased leaf length by 71.3%, 39.2% and 34.8% compared to the non-treated control, addition of manure and DAP alone and sole application of ECO-T respectively. Combining of Rhizatech with fertilizers also significantly affected leaf length recording an increase of 45.8% (Rhizatech plus MRPCAN), 40.8% (Rhizatech plus manure plus DAP) and 48.3% (Rhizatech plus Mavuno) above the sole fertilizer applications. Effect of combining Rhizatech and Mavuno on leaf width was statistically different from all other treatments recording an increase of 100%, 35.9% and 96% compared to the non-treated control and sole application of Mavuno and Rhizatech respectively. Other treatments with effects on leaf length statistically different from the non-treated control include Rhizatech plus MRPCAN, ECO-T plus manure plus DAP, Mavuno alone, ECO-T plus

Mavuno and Rhizatech plus Mavuno increasing leaf length by 55.7%, 62.6%, 55.6%, 52.7% and 100% respectively (Table 4).

Table 4. Growth of tissue cultured banana plants treated with microbiological products and different fertilizers in Rhodic Ferralsol, 22 weeks after deflasking.

Product	Fertilizer	Height	NOL	Girth	LL	LW
None	None	8.0a	7.7a	1.7a	15.6a	5.6a
ECOT	None	10.9b	7.8a	1.6a	19.8b	7.1abc
Rhizatech	None	8.7a	8.2ab	1.6a	16.9a	5.7a
None	MRPCAN	11.8b	9.1c	1.8a	19.5b	6.8ab
ECOT	MRPCAN	11.0b	8.7bc	2.6b	20.1b	7.1abc
Rhizatech	MRPCAN	13.7c	9.9d	1. 9ab	24.6c	8.7cd
None	Manure + DAP	13.6c	7.7a	1.84ab	19.2b	6.9abc
ECOT	Manure + DAP	16.8e	8.6bc	2.1ab	26.7d	9.1d
Rhizatech	Manure + DAP	15.3d	8.9bc	2.0ab	23.6c	6.9abc
None	Mavuno	18.4f	8.7bc	2.2ab	24.3c	8.7cd
ECO-T	Mavuno	16.6de	7.8a	2.1ab	23.8	8.5bcd
Rhizatech	Mavuno	18.9f	8.4abc	2.1ab	25.0cd	11.1e
SED						
Product		0.197	0.112	0.115	0.314	0.275
Fertilizer		0.227	0.129	0.133	0.363	0.318
Product × Fertilizer		0.393	0.224	0.23	0.628	0.551
p value						
Product		<0.001	<0.001	0.147	<0.001	<0.001
Fertilizer		<0.001	<0.001	<0.001	<0.001	<0.001
Product × Fertilizer		<0.001	<0.001	0.068	<0.001	<0.001

Means of 16 plants replicated three times. Within the same column, means followed by the same letter are not significantly different (Tukey's HSD test) at $p > 0.05$. NOL, number of leaves; LL, leaf length; LW, leaf width. Leaf length and leaf width measured in centimeters.

Plants treated with manure alone in the Rhodic Ferralsol had significantly higher number of secondary roots (27) compared to the plants that received neither fertilizer nor product (un-treated control) and plants treated with Rhizatech (Table 5). Effects of all other treatments were statistically the same as the untreated control and Rhizatech. Combined application of fertilizer and products had no significant effects on shoot dry weights (SDWT) in the Rhodic Ferralsol. The highest SDWT (6.07 g) in the Rhodic Ferralsol were observed on plants treated with a combination of Rhizatech plus manure plus DAP and the least on untreated control plants.

Plants treated with Rhizatech alone had significantly lower root dry weights (RDWT) (1.22 g) than plants treated with ECO-T plus manure plus DAP (6.11 g) and Mavuno alone (6.27 g) and statistically the same as all other treatments.

Table 5. Shoot and root growth of tissue cultured banana plants treated with microbiological products and different fertilizers in Rhodic Ferralsol, 22 weeks after deflasking.

Product	Fertilizer	SDWT (g)	RDWT (g)	NOSR
None	None	2.1	1.8ab	14.3a
ECOT	None	4.3	2.1ab	19.3ab
Rhizatech	None	2.5	1.2a	14.7ab
None	MRPCAN	5.4	3.4ab	21ab
ECO-T	MRPCAN	3.3	3.5ab	19ab
Rhizatech	MRPCAN	3.7	2.8ab	18ab
None	Mavuno	7.3	6.3ab	24ab
ECO-T	Mavuno	6.5	5.1ab	24.3ab
Rhizatech	Mavuno	4.9	2.7ab	18ab
None	Manure + DAP	3.4	4.5ab	27.3b
ECO-T	Manure + DAP	5.9	6.1b	20ab
Rhizatech	Manure + DAP	6.1	5.6ab	24.7ab
SED				
Product		0.740	0.643	1.547
Fertilizer		0.855	0.743	1.786
Product × Fertilizer		1.481	1.287	3.093
p value				
Product		0.598	0.178	0.201
Fertilizer		0.006	<0.001	0.001
Product × Fertilizer		0.147	0.343	0.113

Means of 3 replicates. Within the same column, means followed by the same letter are not significantly different (Tukey's HSD test) at $p > 0.05$. SDWT, shoot dry weight; RDWT, root dry weight; NOSR, number of secondary roots.

3.2. Growth of Tissue Cultured Banana Treated with Microbiological Products and Different Fertilizers in a Vertisol Soil

Under green house conditions, plant growth (height, girth, no. of leaves, leaf length and leaf width) was significantly affected by sole and combined application of products and fertilizers (Table 6). Effect of combined application of ECO-T plus manure plus DAP on plant height was statistically different from all other treatments increasing height by 110.6%, 33.7% and 51.5% compared to the non-treated control and sole application of manure plus DAP and ECO-T, respectively. Similar observations were made on plant girth, with combined application of ECO-T plus manure plus DAP increasing plant girth by 22.6%, 9.3% and 83.9% compared to the non-treated control and sole application of manure plus DAP and ECO-T respectively. Significant differences between treatments were observed on number of leaves with the highest number of leaves (11) observed on plants treated with sole application of MRPCAN. Most treatments negatively affected number of leaves up to a magnitude of 9.5%.

Leaf length of plants treated with ECO-T and manure plus DAP was statistically different from all other treatments. An increase of 51.6%, 25.4% and 25.9% above the non-treated control and sole application of manure plus DAP and ECO-T, respectively was observed. Combined application of Rhizatech and Mavuno, ECO-T plus manure plus DAP, Rhizatech plus manure plus DAP, Rhizatech plus MRPCAN had similar effects but statistically higher leaf length than non-treated control, sole application of Rhizatech and combined application of ECO-T and MRPCAN (Table 6).

Combined application of fertilizer and products had no significant effects on shoot dry weights (SDWT) in the Vertisol (Table 7). The highest SDWT was observed with addition of Mavuno alone (13 g) and the least with application of Rhizatech plus MPRCAN) (3.87 g). The highest RDWT was observed with combined application of ECO-T plus manure plus DAP (7.45) and the least on the un-treated control plants. The number of secondary roots on plants treated with Mavuno fertilizer alone (31) was statistically different from plants treated with ECO-T (20) and mijingu rock phosphate plus CAN (MPRCAN) (20) but statistically the same as all other treatments (Table 7).

3.3. Growth and Performance of Tissue Culture Banana Plants Treated with Microbiological Products and Different Fertilizers in Rhodic Ferralsol under Field Conditions

Combining of microbiological products with fertilizers in Rhodic Ferralsol significantly ($p < 0.001$) increased growth of banana plants (Table 8). Plants treated with either fertilizer alone or inorganic fertilizer plus products performed better than plants treated with products only. Combination of ECO-T with manure plus DAP increased plant height by 33.8% and 31.2% above the sole application of manure plus DAP and ECO-T respectively. The effect of combining Rhizatech and MRPCAN on plant height was statistically the same as that of combining ECO-T and manure plus DAP. An increase of 26.5% and 59.7% above sole application of MRPCAN and Rhizatech respectively was recorded with the combined application of Rhizatech and MRPCAN. Plants treated with Rhizatech alone had statistically lower heights than all other treatments.

Table 6. Growth of tissue cultured banana plants treated with microbiological products and different fertilizers in a Vertisol, 22 weeks after deflasking.

Product	Fertilizer	Height (cm)	Girth (cm)	NOL	LL (cm)	LW (cm)
None	None	11.7a	2.1bcd	10.7bcd	23.6a	9.1a
ECO-T	None	16.3c	1.4a	10.8cd	28.5cd	10.9bcde
Rhizatech	None	13.7b	2.1bcd	10.6abcd	25.2ab	9.4a
None	MRP + CAN	15.1bc	2.0bc	11.2d	26.8bc	9.7ab
ECO-T	MRP + CAN	15.6c	2.4def	10.1ab	26.1b	9.2a
Rhizatech	MRP + CAN	19.0d	2.3cde	10.5abc	29.1d	11.4de
None	Manure + DAP	18.5d	2.4def	10.5abc	28.6cd	10.1abc
ECO-T	Manure + DAP	24.7g	2.6f	10.5abc	35.8f	11.6e
Rhizatech	Manure + DAP	22.4f	2.5ef	10.5abc	33.1e	11.5de
None	Mavuno	19.7de	1.9bc	10.4abc	28.4cd	10.3abcd
ECO-T	Mavuno	20.6e	1.8ab	10.0a	29.7d	11.1cde
Rhizatech	Mavuno	22.5f	2.5ef	10.4abc	33.5e	11.8e
SED						
Product		0.219	0.059	0.099	0.059	0.187
Fertilizer		0.253	0.068	0.115	0.068	0.215
Product × Fertilizer		0.439	0.117	0.199	0.117	0.373
p value						
Product		<0.001	<0.001	0.006	<0.001	<0.001
Fertilizer		<0.001	<0.001	0.001	<0.001	<0.001
Product X Fertilizer		<0.001	<0.001	<0.001	<0.001	<0.001

Means of 16 plants replicated three times. Within the same column, Means followed by the same letter are not significantly different (Tukey's HSD test) at p > 0.05. NOL, number of leaves; LL, leaf length; LW, leaf width; MRPCAN, Minjigu Rock Phosphate + Calcium Ammonium Nitrate.

Table 7. Shoot and root growth of tissue cultured banana plants treated with microbiological products and different fertilizers in a Vertisol, 22 weeks after deflasking.

Product	Fertilizer	SDWT(g)	RDWT(g)	NOSR
None	None	4.2	2.6	2.6
ECOT	None	6.2	4.1	4.1
Rhizatech	None	7.4	3.9	3.9
None	MRPCAN	4.8	3.4	3.5
ECO-T	MRPCAN	7.1	6.6	6.6
Rhizatech	MRPCAN	3.9	3.8	3.8
None	Mavuno	9.2	6.8	6.8
ECO-T	Mavuno	7.7	6.5	6.5
Rhizatech	Mavuno	8.7	5.8	5.8
None	Manure+DAP	5.9	7.1	7.1
ECO-T	Manure+DAP	4.6	7.4	7.4
Rhizatech	Manure+DAP	5.6	5.6	5.6
SED				
Product		1.023	0.676	1.483
Fertilizer		1.181	0.781	1.712
Product X Fertilizer		2.045	1.352	2.966
P value				
Product		0.926	0.11	0.233
Fertilizer		0.041	0.001	0.003
Product X Fertilizer		0.458	0.47	0.023

Means of 3 replicates. Within the same column, means followed by the same letter are not significantly different (Tukey's HSD test) at p > 0.05. SWDT, shoot dry weight; RDWT, root dry weight; NOSR, number of secondary roots; MRPCAN, Minjigu Rock Phosphate + Calcium Ammonium Nitrate.

Similar to the observations on plant height, plants treated with the combination of Rhizatech and MRPCAN increased girth by 14% and 27%, respectively compared to sole application of MRPCAN and manure plus DAP, respectively. The highest number of leaves (12) was observed on the plants treated with Rhizatech or ECO-T alone and this was statistically different from MRPCAN and ECO-T plus Mavuno-treated plants, which had the least number of leaves (10). Leaf length was significantly increased by combined application of Rhizatech and MRPCAN (32.7%) and ECO-T plus manure and DAP (21%). All treatments had statistically the same effects on leaf width. The trend of plant growth in the Rhodic Ferralsol is shown in Figure 2.

3.4. Growth and Performance of Tissue Cultured Plants Treated with Microbiological Products and Different Fertilizers in Vertisol under Field Conditions

Plant height in the Vertisol was significantly ($p = 0.021$) affected by the combined application of commercial products with either inorganic or organic fertilizers in (Table 9). Combining of ECO-T with macro and micronutrients (Mavuno) increased growth of plants by 13.5% compared to the sole addition of Mavuno and by 2% compared to the addition of ECO-T alone. Combined application of Rhizatech and with P and inorganic N increased plant height by 11.8% compared to sole addition of P and inorganic N and reduced plant height by 7% compared to sole addition of Rhizatech. The trend of plant growth in the Vertisol is shown in Figure 3.

Plant girth was enhanced by application of Rhizatech plus MRPCAN by 5.1%, 37.2% and 16.1% compared to sole application of Rhizatech, MRPCAN and manure plus DAP respectively. This was followed by combined application of ECO-T plus Mavuno that increased plant girth by 4.8%, 13.8% and 15.8% respectively compared to the sole application of ECO-T, Mavuno or manure plus DAP respectively.

Leaf width was more affected by combined application of Rhizatech and MRPCAN and ECO-T plus manure plus DAP than leaf width (the effect of the two treatments was statistically the same). Combined application of Rhizatech plus MRPCAN increased leaf length by 17%, 35% and 20.8% compared to the sole application of Rhizatech, RPCAN and manure plus DAP respectively. ECO-T plus Mavuno increased leaf length by 14%, 13.6% and 17% compared to the sole application of ECO-T, Mavuno and manure plus DAP.

Table 8. Growth of tissue cultured banana plants treated with microbiological products and different fertilizers, 12 months after field establishment in a Rhodic Ferralsol.

Product	Fertilizer	Height (cm)	Leaf length (cm)	Leaf width (cm)	No. of leaves	Girth (cm)
Rhizatech	None	166.5a	182.6a	63.9	12.bc	18.76
ECO-T	None	205.1ab	204.7ab	63.9	12c	20.8
None	Manure + DAP	206.6ab	203ab	65.1	11abc	18.2
ECO-T	Mavuno	226.7bc	217.8ab	65.1	10a	21.4
None	MRP + CAN	235.8bc	219.2ab	66.4	10a	20.6
None	Mavuno	245.3bc	230.6bc	72.8	11abc	20.9
Rhizatech	MRP + CAN	261.6c	242.4bc	74.5	11abc	23.5
ECO-T	Manure + DAP	273.7c	247.9bc	71.6	116abc	23.2
p value						
Product		0.116	0.193	0.685	0.206	0.160
Fertilizer		0.171	0.496	0.359	0.022	0.287
Product × Fertilizer		0.001	0.019	0.111	0.001	0.146

Means of four replicates. Within the same column, means followed by the same letter are not significantly different (Tukey's HSD test) at p > 0.05. MRPCAN, Minjigu Rock Phosphate + Calcium Ammonium Nitrate.

Table 9. Growth and performance of tissue cultured plants treated with microbiological products and different fertilizers in Vertisol, 12 months after field establishment.

Product	Fertilizer	Height (cm)	Leaf length (cm)	Leaf width (cm)	No. of leaves	Girth (cm)
Rhizatech	None	168.6ab	159.9ab	50.4a	6.6abc	20.6a
EcoT	None	207.3ab	195.8b	62.3b	7.8c	20.1a
None	MRP + CAN	156.7a	140.8a	50.8a	6ab	15.8b
None	Manure + DAP	186.5ab	173ab	58.8ab	6.3abc	18.7a
None	Mavuno	186.2ab	165.6ab	58.6ab	6.1abc	19a
Rhizatech	MRP + CAN	156.7ab	179.5ab	58.9ab	5.5a	17.1ab
Rhizatech	Mavuno	183.1ab	167.6ab	58.9ab	6.7abc	17. ab
ECO-T	Mavuno	211.3b	195.1b	63.1b	7.1bc	21.6a
P value						
Product		0.4	0.3097	0.8957	0.2189	0.169
Fertilizer		0.6896	0.6638	0.592	0.9501	0.7413
Product × Fertilizer		0.0157	<0.0001	0.105	0.003	<0.0001

Means of four replicates. Within the same column, means followed by the same letter are not significantly different (Tukey's HSD test) at p > 0.05. MRPCAN, Minjigu Rock Phosphate + Calcium Ammonium Nitrate.

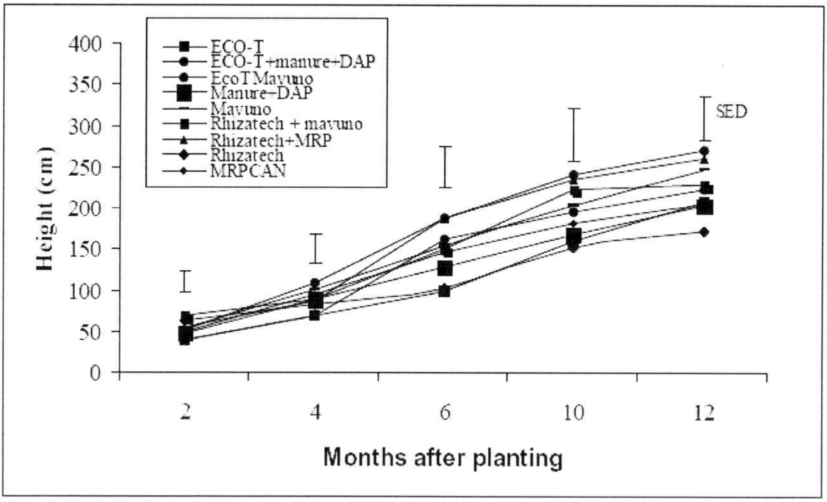

Figure 2. Growth of tissue cultured plants treated with microbiological products and different fertilizers in Rhodic Ferralsol. SED, standard error of difference between means at 12 months after field establishment.

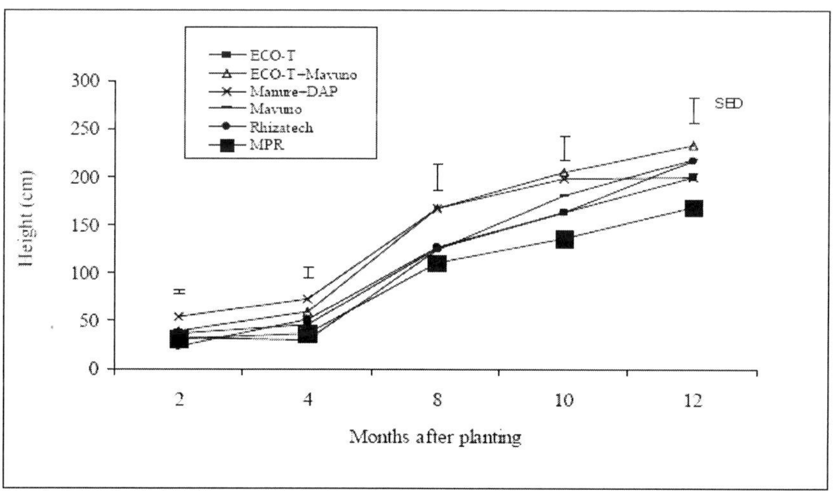

Figure 3. Growth of tissue cultured plants treated with microbiological products and different fertilizers in a Vertisol. SED, standard error of difference between means at at 12 months after field establishment.

4. DISCUSSION

The performance of tissue cultured banana plants treated with combined application of fertilizers and commercial microbiological products was variable and depended on prevailing soil conditions. Results indicate preference for certain sources of inorganic or organic nutrients by the Trichoderma based product (ECO-T) and the mycorrhiza-based product (Rhizatech). The positive effects of combining Trichoderma with organic and inorganic fertilizers were more evident in the Vertisol than that in the Rhodic Ferralsol suggesting a preference for soils with higher nitrogen levels i.e. 0.25% in the Vertisol versus 0.1% in the Rhodic Ferralsol. The performance of plants was greatly enhanced with Trichoderma combined with either organic manure inorganic phosphate diammonium phosphate) or combined with a macro and micro nutrient source (Mavuno) as compared to the sole application of Trichoderma. This indicates that organic and inorganic amendments could enhance the proliferation and efficacy of Trichoderma and subsequently enhance plant growth. Higher amounts of Trichoderma in soils with organic amendments have been reported [17]. The enhanced performance of plants when inoculated with Trichoderma combined with mineral nutrients (macro and micro nutrients) could be attributed to availability of nutrients for plant uptake as well as increased proliferation of Trichoderma populations with a consequent result of improved plant growth. Mineral nutrition is essential for growth, sporulation and stimulation of fungal secondary metabolism [18] [20] as well high total N availability [20]. A positive correlation between soil nitrate levels and cellulose production has also been reported [32]. Performance of plants treated with combination of mycorrhiza and either Mavuno (macro and micro nutrients) and minjigu rock phosphate was consistently better in the Rhodic Ferralsol than either mycorrhiza alone or fertilizer alone. This could suggest that the efficacy of mycorrhiza is enhanced by addition of mineral nutrients to nutrient poor soils, as is the case with the Rhodic Ferralsol [33]. This could also suggest an improved fertilizer use when mycorrhiza is combined with mineral fertilizer. This finding is in agreement with the finding of other authors who have reported that arbuscular mycorrhizal symbiosis provides numerous services to crops [34], including efficient use of fertilizer and soil nutrients [23], increased N-fixation in legumes [35], improving plant access to mineral nutrients, including N, P, K, Ca, Mg, Fe, Cu, Zn, and Mn [36]. Evaluation of the effect of biofertilizers combined with different soil amendments on potted rice plants has been reported to enhance 100-grain weight in the NPK fertilizer amendment [37].

Amendment of both the Vertisol and Rhodic Ferralsol soils with minjigu rock phosphate alone had the lowest effect on plant growth enhancement suggesting minimal solubilization and utilization of rock phosphate by plants. However, plant growth was greatly enhanced with combined application of minjigu rock phosphate and mycorrhiza (Rhizatech) indicating a positive effect of mycorrhiza in the added phosphate (P) solubilization and acquisition of by plants with a consequent increase in plant growth. This finding conforms with the reports of other authors, who have shown that mycorrhiza play an important role in effecting the availability of soil P to plant roots, and increasing P mobilization in soil [38] [39].

5. CONCLUSION

In conclusion, it is evident that tissue culture banana plants could greatly benefit from combined application of commercial microbiological and fertilizers. Combinations of mycorrhiza and inorganic macro and micro nutrients or rock phosphate and the combined application of Trichoderma and organic sources could greatly enhance plant growth in nutrient poor soils as was observed in the Rhodic Ferralsol and Vertisol. Knowledge of product formulations and prevailing soil conditions is essential prior to inoculation and amendment of the soils.

ACKNOWLEDGEMENTS

This work was supported by the Bill and Melinda Gates foundation through the Tropical Soil Biology Institute of CIAT (CIAT-TSBF) project on commercial microbial products. We greatly appreciate the assistance of Edwin Mutegi, Laban Nyabenga and Thomas Ngotho for field technical assistance.

REFERENCES

1. Mink, G.I. (1991) Controlling Virus Diseases Using Disease-Free Stocks. In: Pimental, D., Ed., Handbook of Pest Management in Agriculture, 2nd Edition, CRC Press, Boca Raton, 1, 363-391.

2. Mbaka, J.N., Mwangi, M. and Mwangi, M.N. (2008) Banana Farming as a Business: The Role of Tissue Culture Planting Materials. Journal of Applied Biosciences 9, 354-361. www.biosciences.elewa.org

3. Cassalls, A.C. and O'Herlihy, E.A. (2003) Disease Management of Micropropagated Plants, Good Laboratory Practice. Acta Horticulturae, 616, 105-114.

4. Abdullah, M.Y., Hassan, Z.M. and Talib Z. (1999) Trend in Foliar Nutrient Concentrations and Contents and Its Implication on Leaf Area Index Development and Yield in Banana Cultivar "Berangan". In: Zakaria, W., Mahmud, T.M.M., Siti Khalijah, D., Nor Aini, M.F. and Marziah, M., Eds., Proceedings of First National Banana Seminar, Awana, 95-105.

5. Robinson, J.C. (1996) Bananas and Plantains. CAB International, Wallinford, 172-174.

6. Khan, S.A., Mulvaney, R.L., Ellsworth, T.R. and Boast, C.W. (2007) The Myth of Nitrogen Fertilization for Soil Carbon Sequestration. Journal of Environmental Quality, 36, 1821-1832.

7. Ladha, J.K., de Brujin, F.J. and Malik, K.A. (1997) Introduction Assessing Opportunities for Nitrogen Fixation in Rice a Frontier Project. Plant and Soil, 194, 1-10.

8. Ladha, J.K. and Reddy, P.M. (1995) Extension of Nitrogen Fixation to Rice: Necessity and Possibilities. Geo Journal, 35, 363-372.

9. Mia, M.A. and Shamsuddin, Z.H. (2010) Rhizobium as a Crop Enhancer and Biofertilizer for Increased Cereal Production. African Journal of Biotechnology, 9, 6001-6009.

10. Kennedy, I.R., Choudhury, A.T.M.A. and Kecskés, M.L. (2004) Non-Symbiotic Bacterial Diazotrophs in Crop-Farming Systems: Can Their Potential for Plant Growth Promotion Be Better Exploited? Soil Biolology and Biochemistry, 36, 1229-1244.

11. Jeyabal, A. and Kuppuswamy, G. (2001) Recycling of Organic Wastes for the Production of Vermicompost and Its Response in Rice-Legume Cropping System and Soil Fertility. Europian Journal of Agronomy, 15, 153-170.

12. Reino, J.L., Guerrero, R.F., Hernandez-Galan, R. and Collado, I.G. (2008) Secondary Metabolites from Species of the Biocontrol Agent Trichoderma. Phytochemistry Review, 7, 89-123.

13. Rodriguez-Romero, A.S., Guerra, M.S. and Jaizme-Vega, M.D. (2005) Effect of Arbuscular Mycorrhizal Fungi and Rhizobacteria on Banana Growth and Nutrition. Agronomy for Sustainable Development, 25, 395-399.

14. Benítez, T., Rincón, A.M., Limón, M.C. and Codón A.C. (2004) Biocontrol Mechanisms of Trichoderma Strains. International Journal of Microbiology, 7, 249-60.

15. Harman, G.E., Howell, C.R., Viterbo, A., Chet, I. and Lorito, M. (2004) Trichoderma Species Opportunistic, Avirulent Plant Symbionts. Nature Reviews Microbiology, 2, 43-56.

16. Jaizme-Vega, M.C., Sosa Hernández, B. and Hernández Hernández, J.M. (1998) Interaction of Arbuscular Mycorrhizal Fungi and the Soil Pathogen Fusarium oxysporum f. sp. cubense on the First Stages of "Grande Naine" Banana. Acta Horticulturae, 490, 285-295.

17. Okoth, S., Roimen, H., Mutsotso, B., Muya, E. and Okoth, P. (2007) Land Use Systems and Distribution of Trichoderma Species in Embu Region, Kenya. Tropical and Subtropical Agroecosystems, 7, 105-122.

18. Griffin, D.H. (1994) Fungal Physiology. 2nd Edition, Wiley-Liss, New York.

19. Score, A.J. and Palfreyman, J.W. (1994) Biological Control of the Dry Rot Fungus Serpula lacrymans by Trichoderma Species. International Biodeterioration and Biodegradation, 33, 115-128.

20. Fargasova, A. (1992) The Influence of Various Nitrogen Sources on the Growth, Conidiation and Pigmentation Production of the Brown Mutant of Trichoderma viridae M-108. Biology, 47, 453-464.

21. Schubert, A. and Hayman, D.S. (1986) Plant Growth Responses to Vesicular-Arbuscular Mycorrhiza XVI. Effectiveness of Different Endophytes at Different Levels of Soil Phosphate. New Physiologist, 103, 79-90.

22. Marschner, P. and Baumann, K. (2003) Changes in Bacterial Community Structure Induced by Mycorrhizal Colonization in Split-Root Maize. Plant and Soil, 251, 279-289.

23. Javaid, A. (2009) Arbuscular Mycorrhizal Mediated Nutrition in Plants. Journal of Plant Nutrition, 32, 1595-1618.

24. Atul-Nayyar, A., Hamel, C., Hanson, K. and Germida, J. (2009) The Arbuscular Mycorrhizal Symbiosis Links N Mineralization to Plant Demand. Mycorrhiza, 19, 239-246.

25. Bedini, S., Avio, L., Argese, E. and Giovannetti, M. (2007) Effects of Long-Term Land Use on Arbuscular Mycorrhizal Fungi and Glomalin-Related Soil Protein. Agriculture, Ecosystems & Environment, 120, 463-466.

26. Hijri, I., Sykorova, Z., Oehl, F., Ineichen, K., Mäder, P., Wiemken, A. and Redecker, D. (2006) Communities of Arbuscular Mycorrhizal Fungi in Arable Soils Are Not Necessarily Low in Diversity. Molecular Ecolology, 15, 2277-2289.

27. Mäder, P., Fliessbach, A., Dubois, D., Gunst, L., Fried, P. and Niggli, U. (2002) Soil Fertility and Biodiversity in Organic Farming. Science, 296, 1694-1697.

28. Anderson, J.M. and Ingram, J.S. (1993) Tropical Soil Biology and Fertility: Handbook of Methods. CAB International, Wallingford.

29. Okalebo, J.R., Gathna, K.W. and Woomer, P.L. (2002) Laboratory Methods for Soil and Plant Analysis. A Working Manual. 2nd Edition, Tropical Soil Fertility and Biology Program, TSBF-CIAT and SACRED Africa, Nairobi, 128 p.

30. Kavoo-Mwangi, A.M., Kahangi, E.M., Ateka, E., Onguso, J., Mukhongo, R.W., Mwangi, E.K. and Jefwa, J.M. (2013) Growth Effects of Microorganisms Based Commercial Products Inoculated to Tissue Cultured Banana Cultivated in Three Different Soils in Kenya. Applied Soil Ecology, 64, 152-162.

31. SAS Institute Inc. (2006) SAS User's Guide: Statistics. SAS Institute Inc., Cary.

32. Widden, P., Cunningham, J. and Breil, B. (1989) Decomposition of Cotton by Trichoderma Species: Influence of Temperature, Soil Type, and Nitrogen Level. Canadian Journal of Microbiology, 35, 469-473.

33. FAO (2006) World Agriculture towards 2030/50. Interim Report, FAO, Rome.

34. Gianinazzi, S., Gollotte, A., Binet, M.N., van Tuinen, D., Redecker, D. and Wipf, D. (2010) Agroecology: The Key Role of Arbuscular Mycorrhizas in Ecosystem Services. Mycorrhiza, 20, 519-530.

35. Barea, J.M., Azcon, R. and Azcon-Aguilar, C. (1983) Interactions between Phosphate Solubilizing Bacteria and VA Mycorrhiza to Improve Plant Utilization of Rock Phosphate in Non-Acidic Soils. 3rd International Congress on Phosphorus Compounds, Brussels, 4-6 October 1983, 127-144.

36. Liu, A., Hamel, C., Hamilton, R.I., Ma, B.L. and Smith, D.L. (2000) Acquisition of Cu, Zn, Mn, and Fe by Mycorr-Hizal Maize (Zea mays L.) Grown in Soil at Different P and Micronutrient Levels. Mycorrhiza, 9, 331-336.

37. Javaid, A. (2011) Effects of Biofertilizers Combined with Different Soil Amendments on Potted Rice Plants. Chilean Journal of Agricultural Resources, 71, 157-163.

38. Toro, M., Azcón, R. and Barea, J.M. (1997) Improvement of Arbuscular Mycorrhiza Development by Inoculation of Soil with Phosphate-Solubilizing Rhizobacteria to Improve Rock Phosphate Bioavailability (32P) and Nutrient Cycling. Applied and Environmental Microbiology, 63, 4408-4412.

39. Smith, S.E. and Read, D.J. (2008) Mycorrhizal Symbiosis. 3rd Edition, Elsevier Academic Press, Amsterdam, 787.

Chapter 6

Role of Pigeonpea Cultivation on Soil Fertility and Farming System Sustainability in Ghana

S. Adjei-Nsiah

Forest and Horticultural Crops Research Centre, Kade, Institute of Agricultural Research, College of Agriculture and Consumer Sciences, University of Ghana, P.O. Box 68, Legon, Ghana

ABSTRACT

The productivity of the smallholder farming system in Ghana is under threat due to soil fertility decline. Mineral fertilizer is sparingly being used by smallholder farmers because of prohibitive cost. Grain legumes such as pigeonpea can play a complementary or alternative role as a source of organic fertilizer due to its ability to enhance soil fertility. Despite its importance, the potential of pigeonpea as a soil fertility improvement crop has not been exploited to any appreciable extent and the amount of land cultivated to pigeonpea in Ghana is vey negligible. This paper synthesizes recent studies that have been carried out on pigeonpea in Ghana and discusses the role of pigeonpea cultivation in soil fertility management and its implication for farming system sustainability. The paper shows that recent field studies conducted in both the semi-deciduous forest and the forest/savanna transitional agro-ecological zones of Ghana indicate that pigeonpea/maize rotations can increase maize yield by 75–200%. Barrier to widespread adoption of pigeonpea include land tenure, market, and accessibility to early maturing and high yielding varieties. The paper concludes among other things that in order to promote the cultivation of pigeonpea in Ghana, there is the need to introduce varieties that combine early maturity with high yields and other desirable traits based on farmers preferences.

1. INTRODUCTION

Agricultural productivity in the smallholder farming systems in Ghana is under threat due to declining soil fertility. In the past, smallholder farmers in Ghana relied on the extended bush fallow system for maintaining the productivity of their farmlands [1]. This system allowed restoration of phosphorus (P) and nitrogen (N), the most limiting nutrients. However, over the years, the population growth-induced scarcity of suitable farmland has led to the shortening of the fallow period making it difficult to manage soil fertility in smallholder farming systems. The problem is compounded by the increasing cost of inputs at the farm level due to structural adjustment programmes that have removed subsidies and increased supply costs due to the deterioration conditions of rural infrastructure [2]. For instance, in 2002, whereas a metric tonne of urea cost about US$90 FOB (free on board) in Europe [3], the same quantity cost a Ghanaian farmer about US$308 at the farm level [4].

Most farmers, especially the smallholder farmers, do not have access to formal credit and therefore cannot afford to buy mineral fertilizers even when it has been demonstrated to be profitable [5]. Furthermore, due to the unpredictability of rainfall in some areas, farmers consider fertilizer application as risky.

In smallholder farming systems in Ghana, legumes can play a complementary or alternative role as source of organic fertilizer. Research in many parts of Sub-Saharan Africa including Ghana has shown that legumes have the potential to sustain soil fertility in smallholder farming systems [6–8]. Pigeonpea (Cajanus cajan (L.) Millsp.) has been found to have a great potential in this respect because of its ability to recycle nutrients and tolerate wide environmental conditions and low soil fertility [9]. Pigeonpea is nutritionally well balanced and is an excellent source of proteins (20–30%) [7]. In addition to proteins, pigeonpea provides carbohydrates and high levels of vitamins A and C.

Despite the importance of pigeonpea in terms of nutrition and soil fertility improvement, the crop has not been promoted to any appreciable extent by the national agricultural research and extension system in Ghana. In the forest/savannah transitional zone where the crop is widely cultivated, pigeonpea is considered as women's crop and therefore minor due to men's dominion over women in household decision-making process. Furthermore the overemphasis on maize in the current production systems as the major food security and cash crop in the forest/savanna transitional zone has relegated pigeonpea in particular and legumes in general as a minor crop.

However, recent studies in the forest/savannah transitional agroecological zone as well as the semideciduous forest zone of Ghana have demonstrated that when integrated in the cropping system as a form of rotation, pigeonpea has the potential of improving soil fertility. This paper discusses the role of pigeonpea cultivation in soil fertility management and its implication for farming system sustainability and food security.

2. MATERIAL AND METHODS

This paper is largely based on studies that have been conducted in the forest/savannah transitional agro-ecological zone of Ghana between 2002 and 2006 under the framework of Convergence of Science project [10] and between 2007 and 2010 under the IDRC funded Climate Change Adaption programme in Africa [11]. Results from the 2002–2006 studies have already been published [8, 9, 12], while results from 2007–2010 studies have not been published.

2.1. The Study Area and Population

The studies were conducted in the Wenchi Municipal (7° 27 and 8° 30 N, 1° and 2° 36 W) in the forest/savanna transitional agro-ecological zone of Ghana. The relief of Wenchi is gently undulating to flat. The soils, which are mainly Lixisols, are fragile with shallow top soils underlain with compact concretions and impermeable iron pans [8]. Temperatures are relatively high with a monthly mean of about 30°C. Rainfall is bimodal and starts in April and ends in November with a dry spell in August. The rainy season is followed by a long dry season from November to March. The annual rainfall is about 1,300 mm with about 107 rainy days. Wenchi municipal, which has a total population of 97,058 (2000 census), is ethnically diverse with about 20% of the population being migrants from the three northern regions of Ghana and the neighbouring Burkina Faso.

2.2. Research Approach and Methods

In 2002, we carried out a study under the Convergence of Sciences project to explore farmers' soil fertility management strategies and their relevant social context [8]. In order to ground the research in the needs of the farming communities, a diagnostic study was carried out in the study area between July, 2002 and July, 2003 using Participatory Rural Appraisal tools such as drawing of a community territory map (to identify the differences in soil fertility patterns), a transect walk (to reveal the diversity of the landscape), and analysis of soil fertility management strategies and group discussions. Group discussions with 10–40 farmers were held in the village centre and/or on farmers' fields.

In addition, two sets of individual interviews with farmers were conducted to collect qualitative and quantitative data. In the first interview, which involved 40 farmers, the selection of farmers was done through stratified sampling. A list of farmers in the community was obtained from the village committee secretary and every tenth name from the list was selected for individual interviewing. The second interview which involved 38 farmers was conducted later to look at the farming characteristics of the various subcommunities in the village using a wealth ranking exercise. For this interview, 6–10 persons were selected from each wealth category within each subcommunity. The individual interviews were semistructured in nature and served both to get more quantitative data on farm size, household composition, and the farming system and to obtain a better qualitative understanding of the soil fertility management strategies and their underlying rationale.

The diagnostic study was followed by farmer participatory on-farm experimentations with three farmer research groups established soon after the diagnostic study to evaluate the agronomic efficacy of the soil fertility management practices being used by the farmers [9]. Six cropping sequences: cassava cropping; pigeonpea cropping; mucuna/maize/mucuna rotation; cowpea/maize/cowpea rotation; maize/maize/maize; Imperata cylindrica fallow were evaluated on both farmer-managed and researcher-managed plots for their effects on soil fertility and yield of subsequent maize test crop. To deepen our understanding of soil fertility management, we carried out further exploration of diversity among the farmers according to gender, ethnicity, and wealth [12]. Farmers were selected from three communities in Wenchi according to ethnicity and gender for interview using semistructured questionnaires. We conducted two sets of interviews. For the first interview, the native households were categorized into male-headed and female-headed households. Subsequently, a stratified sample was selected consisting of 20 males from male-headed households, 20 females from male-headed households, and 20 females from female-headed households. In the case of the migrants, every farmer in the community was interviewed because of the small size of their population. As migrant women do not have their own farming enterprises, only males were interviewed. In the second interview, the farmers were selected through a wealth ranking exercise. Fifteen farmers were selected from each of three wealth categories for interviewing. In addition, focus group discussions were held with chiefs, community leaders, family heads, and opinion leaders about land tenure systems in Wenchi.

In 2008, under the IDRC, funded climate change adaptation in Africa project, farmers evaluated three early maturing and three late maturing pigeonpea varieties obtained from the International Crop Research Institute of the Semi-Arid Tropics (ICRISAT) in India, in a community in Wenchi called Asuoano.

3. RESULTS AND DISCUSSION

3.1. Farmers Views on Soil Fertility Decline and Their Causes

Farmers used several terms to express the fertility status of their soils although the terminology differed from one ethnic group to another. For instance, among the Akans and the Walas, when the soil loses its fertility, the farmers would say the soil is "tired" and therefore must be allowed to "rest" under bush fallow to regain its lost "energy." Farmers used various indicators such as colour, water holding capacity, and soil texture to assess soil fertility. A black soil was considered fertile while gravelly and sandy soils were considered less fertile. Other indicators used by farmers to express soil fertility included presence of earthworm casts which the farmers call "earth worm faeces," growth of crops, decline in crop yield and proliferation of certain plant species and weeds. For instance, the presence of weeds like chromolaena odorata and elephant grass (Pennisetum purpureum) indicated a fertile soil while the presence of weeds like spear grass (Imperata cylindrica) indicated a less fertile soil.

Farmers pointed out five major factors as being responsible for soil fertility decline in the area as follows: (1) increasing population as a result of migration of people into the area leading to smaller farms, which has resulted in the continuous cropping of the same piece of land over a long period; (2) annual bushfires which destroy vegetation and the population of earthworms that contribute to soil fertility improvement; (3) monocropping of maize without rotating it with any other crop; (4) the rapid increase in the monetary value of land; (5) felling of trees which exposes the land to the direct action of the sun.

3.2. Commonly Grown Crops in the Area

Table 1 shows the major food crops and grain legumes grown in the area. In terms of magnitude, maize was the most important food crop grown in Asuano. Other food crops included cassava, yams, cocoyam, and plantain. Legumes also played important role in the cropping systems in the area. Farmers listed about seven different types of legumes they cultivated (Table 1). In terms of proportion of land allocated to legumes, farmers ranked cowpea as the most important legume, followed by groundnut and pigeonpea. However, the type of legumes cultivated by farmers differed among the natives and the migrants. While the native farmers preferred pigeonpea, the migrant farmers preferred cowpea and groundnut. This was related to dynamics around land tenure. In Wenchi area, migrants do not own land but access land for farming through share cropping and land renting arrangements and hence preferred short duration legumes like cowpea and groundnut. Long duration legume such as pigeonpea was therefore not a preferred option for migrant farmers. Moreover, the market for pigeonpea is not well developed as it is a local food crop. It must be noted that the sum of the percentage number of people cultivating the various food crops added up to more than 100% because some farmers cultivated more than one crop since smallholder farmers in Ghana in general and Wenchi in particular practise mixed cropping. Yam was planted on mounds to increase the volume of soil for the development of tubers. Maize, cocoyam, and cassava were usually planted on the lower portions of the yam mounds while pigeonpea was planted in the rows between the yam mounds in a sequential order for efficient utilization of nutrients. Yam was usually planted first followed by maize, cocoyam, cassava, and lastly pigeonpea. The legumes mainly cowpea, groundnuts, and pigeonpea were grown to serve several purposes. These included (1) household food security; (2) household income; (3) soil fertility improvement through build up of soil organic matter and release of nutrients, particularly N for subsequent crops. Where crop rotation involving these legumes and maize was practised, crop succession was planned such that maize was planted after harvesting the legume to take advantage of the nitrogen fixed by the legume. In this way, it helped to either reduce or avoid the use of nitrogen fertilizers which may increase the cost of production.

3.3. Role of Pigeonpea in the Cropping Systems in Wenchi

Among the legumes cultivated by the natives, pigeonpea was grown on a larger scale in comparison with other legumes because of its ability to regenerate soil fertility, its low production cost, its tolerance to pests and diseases, its cash income, and its food value. It was generally grown on less fertile land and land with problematic weeds such as spear grass (Imperata cylindrica).

After cropping a piece of land to crops like maize, cassava, and yam for about three to four years, farmers intercropped their food crops with pigeonpea during the last cropping year of the cycle. After harvesting the component crops, farmers allowed the land to remain under pigeonpea for 18–24 months after which the pigeonpea plants were cut down, burnt and the land cropped to maize or yams. The pigeonpea canopy was perceived to protect the soil from the direct action of the sun and therefore prevents the soil from becoming hardened. According to the farmers, pigeonpea forms a canopy after one year and shades out obnoxious weeds by suppressing their growth. The farmers also explained that the leaf litter covers the soil, reduces soil erosion, improves infiltration, prevents heating of the soil, and enhances earthworm activity. Crops grown on the land after pigeonpea, and especially maize, were perceived by the farmers to look greener, grow faster, and yield more. Rotations involving pigeonpea was often regarded by farmers as a long-term soil fertility management strategy. Despites its importance, migrant farmers preferred to use mainly short-term strategies such as mounding and the planting of short duration leguminous crops such as cowpea and groundnut for maintaining soil fertility because of the dynamics surrounding land tenure.

Table 1. The most important food crops and proportion of land allocated to leguminous crops by native and migrant farmers in Asuoano in 2002.

	Natives	Migrants
	%	
Food crop		
Maize	100	100
Cassava	100	88
Yam	100	94
Cocoyam	91	19
Plantain	32	0
Leguminous crop		
Pigeonpea	50	0
Groundnut	20	35
Cowpea	10	50
White kidney bean	10	0
Bambara groundnut	5	15
Mucuna	2	0
Others	3	0

Source: [8].

3.4. Agronomic Potential of Pigeonpea

Studies in Wenchi and Kade in the forest/savanna transitional and semideciduous forest zones of Ghana, respectively, indicate the potential of pigeonpea in improving soil fertility and farm profitability (Tables 2 and3). In the transitional zone, rotating pigeonpea with maize resulted in about 75% increase in maize yield over the continuous maize control. In the semi-deciduous forest zone of Ghana, including pigeonpea in the cropping system also resulted in about 100–200% increase in maize grain yield over continuous maize. In the transitional zone, pigeonpea yielded about 25.5 tons of shoot biomass within 16 months of crop growth. High amounts of leaf litters of up to about 2 t ha^{-1} in one season could be considered a potential source of nutrients for subsequent crop as confirmed by a 75% increase in maize yield in the transitional zone and up to 200% increase in maize yield in the semi-deciduous forest zone of Ghana.

Higher maize yield under legume-maize rotation is expected since legumes are known to fix nitrogen, thereby improving the soil nitrogen economy and enhancing the growth of subsequent crop [6, 14, 15]. The higher maize grain yield associated with pigeonpea could largely be attributed to the large amounts of recycled N incorporated into the soil just before planting the maize test crops. The vigorous root system of pigeonpea has the capacity to explore large soil volume and recycles nutrients from deeper soil profiles. The faster decomposition of the biomass of pigeonpea and N release compared with that of maize stover was better synchronized with maize demand than the slower release of N by the poorer quality materials like maize stover. The N release patterns of organic residues of differential quality have been discussed extensively by [16, 17]. According to [16, 17], poor quality organic residues, with high C : N ratio, provide abundant supply of C for microbial growth leading to immobilization of soil N in the microbial biomass. Residues of high quality organic inputs on the other hand decompose quickly and may release about 70% of the N within a season under tropical conditions [18].

Besides the beneficial effect of pigeonpea on subsequent maize crops as a result of the high N recycling properties of the pigeonpea, the role of other nutrients cannot be ruled out. In some studies, increased soil P availability under pigeonpea was attributed to the efficient solubilisation and uptake of P from bound sources (e.g., Fe-P) by root exudates [19]. Benefit/cost analysis of different cropping sequences involving three grain legumes and cassava over four cropping seasons indicated that pigeonpea/maize rotations had 108% return on investment compared with 31% by continuous maize when fertilizer was not applied to the maize crop (Table4).

Table 2. Effect of crop sequence and N rate on (a) maize grain yield (kg ha^{-1}) and (b) weed biomass (kg ha^{-1}) associated with the maize crop at 8 weeks after planting on researcher-managed plots.

(a)

Crop sequence	N rate (kg ha^{-1})		Mean
	O	60	
Speargrass fallow	1050	2848	1949
Cassava	3002	2738	2870
Pigeonpea	2422	2972	2697
Cowpea-maize-cowpea	1670	2328	1999
Mucuna-maize-mucuna	2970	4195	3582
Maize-maize-maize	1380	2128	1754
Mean	2082	2868	

SED: crop sequence (CS) = 318.4; N rate (NR) = 115.3; CS × NR = 375.8.
P < F: CS = 0.001; NR = 0.001; CS × NR = 0.01.
Source [9].

(b)

Crop sequence	N rate (kg ha^{-1})		Mean
	O	60	
Speargrass fallow	585	790	686
Cassava	270	300	285
Pigeonpea	390	500	445
Cowpea-maize-cowpea	325	395	360
Mucuna-maize-mucuna	300	345	323
Maize-maize-maize	240	430	335
Mean	351	460	

SED: crop sequence (CS) = 65.9; N rate (NR) = 52.4; CS × NR = 112.1.
P < F: CS = 0.001; NR = 0.05; CS × NR = NS.
Source [9].

Table 3. Effect of crop sequence on maize grain yield at 12% moisture content.

Crop sequence	Maize grain yield (t ha^{-1})	
	2007/2008	2008/2009
Pigeonpea	7.0	2.3
Ex-Subi	4.6	1.9
Boakentemma	3.5	1.4
Cowpea-Maize-Cowpea	3.0	1.5
Groundnut-Maize-Groundnut	2.7	1.6
Maize-Maize-Maize	2.3	1.3
LSD at 5%	1.4	0.58

Source [13].

3.5. Challenges in the Cultivation of Pigeonpea

Barrier to widespread adoption of pigeonpea include land tenure, market, and availability of early maturing varieties. In the transitional zone of Ghana where the crop is widely cultivated by farmers, migrant farmers prefer cowpea cultivation to pigeonpea cultivation although they acknowledge the superiority of the latter over the former in terms of soil fertility management and the yield of subsequent maize crop (Table 5). According to the migrants, the market for pigeonpea is not always readily available and return on investment is too slow. In addition, they were afraid that when they invest in the soil, they would not be allowed to reap the full benefit. Among the native farmers, female and male farmers differed in the cultivation of legumes such as pigeonpea and cowpea. Women farmers preferred pigeonpea over cowpea due to its role in food security as well as its low labour requirement [12]. Male farmers preferred cowpea over pigeonpea because the market for cowpea is developed than that of pigeonpea.

Table 4. Estimated costs of production, gross revenue, and returns on investment of (a) various crop sequences (b) maize grown after the sequences with N application to the maize and (c) maize grown after the sequences without N application to the maize.

(a) Crops in the sequence								
[1]Cassava	31,000	2545.1	41.7	41.7	635.0	718.4	1826.7	254
[2]Pigeonpea	1,870	623.3	41.7	8.3	221.5	271.5	351.8	130
[3]Mucuna-maize-mucuna	2,016	365.1	41.7	41.7	247.4	330.8	34.3	10
[4]Cowpea-maize-cowpea	2,536* (1,230)	1079.0	41.7	106,1	475.1	622.9	456.1	73
[5]Maize-maize-maize	3,287	595.2	41.7	36.1	386.1	463.8	456.1	28
[6]Speargrass fallow	0	0	41.7	0	0	41.7	−41.7	−100
(b) Maize after crop sequence with N application								
CS 1	2,738	495.9	13.9	104.2	190.2	308.3	187.6	61
CS 2	2,974	538.5	13.9	104.2	196.5	314.6	223.9	71
CS 3	4,194	759.4	13.9	104.2	245.9	364.0	395.4	108
CS 4	2,331	422.1	13.9	104.2	177.0	295.1	127.0	43
CS 5	2,126	385.0	13.9	104.2	175.4	293.5	91.4	31
CS 6	2,848	515.7	13.9	104.2	224.4	342.5	173.3	51
(c) Maize after crop sequence without N application								
CS 1	3,000	543.2	13.9	13.9	175.6	203.4	339.8	167
CS 2	2,423	438.8	13.9	13.9	165.5	193.3	245.5	127
CS 3	2,961	537.7	13.9	13.9	209.7	237.5	300.2	126
CS 4	1,772	302.8	13.9	13.9	155.2	183.0	119.8	66
CS 5	1,380	249.9	13.9	13.9	153.0	180.7	69.1	38
CS 6	1,048	189.8	13.9	13.9	173.7	200.9	−11.1	−6

[1]US\$82.1 t^{-1}.
[2]US\$333.3 t^{-1}.
[3]US\$181.1 t^{-1} for maize.
[4]US\$337.5 t^{-1} for cowpea and US\$181.1 t^{-1} for maize.
[5]US\$181.1 t^{-1} for maize.
[6]US\$0.
*Yield of maize (figure in brackets).
CS 1: cassava; CS 2: Pigeonpea; CS 3: Mucuna-maize-mucuna; CS 4: Cowpea-maize-cowpea; CS 5: Maize-maize-maize; CS 6: Speargrass fallow.
Source: [9].

Table 5. Preferential ranking of different soil fertility management practices by native and migrant farmers in Wenchi.

Management practice	Ranking order*							
	Natives				Migrants			
	Asuoano[a] N = 10	Beposo[b] N = 5	Droboso[c] N = 7	Average	Asuoano[d] N = 6	Beposo[d] N = 6	Droboso[e] N = 5	Average
(a) Ranking by natives and migrants								
Cassava	1	1	1	1	2	2	1	1.7
Pigeonpea	2	5	2	3	4	4	4	4
Mucuna/Maize/Mucuna	7	6	4	5.7	5	6	6	5.6
Groundnut/Maize/Groundnut	4	3	3	3.3	3	3	3	3
Cowpea/Maize/Cowpea	3	2	5	3.3	1	1	2	1.3
Maize/Maize/Maize	8	7	6	7	7	7	7	7
Cowpea/Cowpea/Cowpea	5	4	7	5.3	6	5	5	5.3
Bush fallow	6	8	8	7.3	8	8	8	8
(b) Ranking by female and male Bonos	Females N = 13	Males N = 10						
Cassava	1	1						
Pigeonpea	2	3						
Mucuna/Maize/Mucuna	5	7						
Groundnut/Maize/Groundnut	3	4						
Cowpea/Maize/Cowpea	4	2						
Maize/Maize/Maize	8	8						
Cowpea/Cowpea/Cowpea	7	5						
Bush fallow	6	6						

[a]Consisted of 6 males and 4 females; [b]Consisted of 4 males and 1 female; [c]Consisted of 6 females and 1male; [d]Dagarbas; [e]Walas.

*Each treatment was compared directly against the other until they were ranked from the highest to the lowest with 1 being the highest ranking and 8 being the lowest.

Source: [9].

Access of farmers to early maturing varieties is also another barrier to widespread adoption of the pigeonpea. Although the late maturing indeterminate pigeonpea varieties have the higher potential to improve soil fertility than the early maturing varieties due to greater biomass production, farmers prefer the early maturing determinate varieties. Criteria farmers used for selecting varieties for planting included maturity, plant height (for ease of harvesting), tolerant to insects, plant architecture (prefer varieties with smaller plant canopy for ease of intercropping with yams), and seed size (bigger seed size is preferred). Amount of biomass and/litter falls and ability to improve soil fertility were not mentioned as criteria although farmers claimed that soil fertility was one of their major production challenges and that pigeonpea was one of the crops they used to improve the fertility of their soils. Although late maturing indeterminate pigeonpea varieties with woody stems are a potential source of fuel wood, farmers did not also include this in their selection criteria. This may be due to the fact that accessibility to fuel wood was not a major constraint in Wenchi.

During focus group discussions in the field, the farmers noted that the late maturing pigeonpea varieties had produced a lot of biomass and had higher litterfalls compared with the early maturing varieties. The farmers observed and commented on the greater litterfall from the late maturing varieties. They further stated that the greater production of biomass and the higher litterfall of the late maturing varieties would lead to better soil fertility improvement compared with the early maturing pigeonpea varieties. Despite their recognition of the soil fertility benefits of the late maturing varieties in improving soil fertility, the farmers did not include this as an important criterion for selecting which pigeonpea varieties to grow (Table 6). Mapfumo et al. [20] made a similar observation in Zimbabwe and suggested that pigeonpea is less likely to be adopted merely on the basis of its capacity to improve soil fertility. Farmers were thus interested in pigeonpea as a food security crop; soil fertility benefits were secondly. Farmers therefore opted for short duration/early maturing varieties which they could easily crop twice in a year. Thus the varieties ICPL 88039 and ICPL 88034 were the most preferred varieties because of their early maturity and high yielding. Although the variety ICPL 87091 was also preferred for their larger grain size, this variety was very susceptible to pod borers.

Table 6. (a) Farmers' preference ranking and (b) farmers' criteria for selection of pigeonpea varieties for planting.

(a) Variety	Growth and yield characteristics					Ranking and reasons for the rank	
	Flowering pattern	Maturity period (days)	100 grains weight (g)	Yield (tons ha^{-1})	Seed colour	Ranking	Reasons for the rank
ICPL 88039	Determinate	125	10.1	1.6	Brownish red	1	Early maturing, short plant height, could be planted twice in a year
ICPL 88034	Determinate	125	10.2	1.8	Brownish red	2	Early maturing, short plant height, could be planted twice in a year
ICPL 87091	Determinate	125	12.1	1.1	white	3	Early maturing, short plant height, could be planted twice in a year
Maruti	Indeterminate	190	10.0	NA	Brownish red	4	Medium maturity, high litter fall
Lakshmi	Indeterminate	196	10.2	NA	Brownish red	4	High litter fall
ICPL 96053	Indeterminate	195	10.1	NA	Brownish red	4	High litter fall

(b) Criteria	Ranking
Maturity	1
Number of plantings in a year	2
Easy of harvesting (plant height)	3
Easy of intercropping with yams	4
Seed size	5

NA: not available.

Source: Adjei-Nsiah (unpublished results).

3.6. Implication of Large Scale Cultivation for System Sustainability

In addition to food uses, pigeonpea has outstanding soil improvement and conservation properties. The growth habit facilitates soil protection, as the canopy continues to expand during the dry season after the component crops in the mixed cropping have been harvested. Living and senesces pigeonpea leaves may offer protective cover for the soil during dry season to prevent the soil from drying out and reduce soil erosion and enhance rain percolation during the rainy season. In farming systems with minimal application of external inputs, management of organic resources plays a major role in maintaining both nutrient availability and soil organic matter [21]. In a cereal-based farming system as that found in the forest/savanna transitional agro-ecological zone of Ghana, where external input use is minimal, most recycling of N and P occurs through pigeonpea litterfall and green leafy biomass of pigeonpea incorporated into the soil after pigeonpea harvest [9]. Pigeonpea litterfall and green leafy biomass of pigeonpea are important sources of easily mineralizable N due to their high nitrogen (1.9 and 2.2% for litterfall and green leafy biomass, resp.) leading to high decomposition rates. Thus, rotation involving pigeonpea within smallholder agriculture has the potential of maintaining a reasonable supply of N and P to cereal crops, particularly maize considering the minimal use of external inputs in a maize-based farming system.

4. CONCLUSION

The paper shows the potential of pigeonpea in improving soil fertility in the smallholder farming systems particularly in a predominantly maize-based farming system as that found in Wenchi, Ghana. We have shown that the potential of pigeonpea in sustaining the productivity of predominantly maize-based farming system is due to its nutrient recycling properties as well as to its role in food security and flexibility in external input use and labour requirement. Soil fertility was apparently not a major criterion during varietal ranking by farmers suggesting that pigeonpea is less likely to be adopted merely on the basis of its capacity to improve soil fertility. Even when there was no strong market demand for pigeonpea, female farmers still integrated pigeonpea in their rotational system. Pigeonpea is largely a women's crop and therefore the attitude by men as major household decision makers has probably undermined its adoption in a predominantly maize based cropping system in Wenchi. Empowering women to make decision with respect to allocation of production resources may enhance a greater adoption of legumes. As more farmers resort to putting their land under pigeonpea instead of bush fallowing, pigeonpea cultivation could serve as an entry point for farming system sustainability. There is, however, the need to (i) introduce short duration pigeonpea varieties that will fit well in the predominantly mixed cropping systems of smallholder agriculture; (ii) evaluate the nutrient recycling capacities of different pigeonpea genotypes; (iii) develop crop rotation/sequencing and soil management options that can improve and/or sustain the productivity of pigeonpea through integrated soil fertility management (ISFM); (iv) design a range of social arrangements that will encourage investment in soil fertility through integration of pigeonpea in the farming system in a heterogeneous farming community like Wenchi, (v) develop use for pigeonpea in new cooking options and/or value added products.

ACKNOWLEDGMENTS

The author is grateful to the two anonymous reviewers who reviewed an earlier version of this paper. Financial support towards this work provided by the Interdisciplinary Research and Education Fund (INREF) of Wageningen University and Research Centre, and The Netherlands and the Dutch Ministry for International Cooperation (DGIS) is gratefully acknowledged. The author also thanks the International Development Centre for funding the climate change adaption in Africa project through the University of Zimbabwe.

REFERENCES

1. P. H. Nye and D. Stephen, “Soil fertility,” in Agriculture and Landuse in Ghana, J. B. Wills, Ed., pp. 127–143, Oxford University Press, London, UK, 1962.

2. H. Gerner, E. O. Asante, E. Owusu Benoah, and K. Marfo, Ghana Fertilizer Privatization Scheme. Private Sector Role and Public Sector

Responsibility in Meeting Needs of Farmers, IFDC-Africa, Lome, Togo, 1995.

3. P. A. Sanchez, "Soil fertility and hunger in Africa," Science, vol. 295, no. 5562, pp. 2019–2020, 2002.

4. ISSER, The state of Ghana economy in 2004, University of Ghana, Legon, Ghana, 2005.

5. H. B. Obeng, K. G. Erbyn, and E. O. Asante, "Fertilizer requirements and Use in Ghana," Consultancy Report, Tropical Agricultural Development Consultancy, Accra, Ghana.

6. K. E. Giller, Nitrogen Fixation in Tropical Cropping Systems, CAB International, Wallingford, UK, 2nd edition, 2001.

7. S. S. Snapp, R. B. Jones, E. M. Minja, J. Rusike, and S. N. Silim, "Pigeon Pea for africa: a versatile vegetable—and more," HortScience, vol. 38, no. 6, pp. 1073–1079, 2003.

8. S. Adjei-Nsiah, C. Leeuwis, K. E. Giller et al., "Land tenure and differential soil fertility management practices among native and migrant farmers in Wenchi, Ghana: implications for interdisciplinary action research," NJAS-Wageningen Journal of Life Sciences, vol. 52, no. 3-4, pp. 331–348, 2004.

9. S. Adjei-Nsiah, T. W. Kuyper, C. Leeuwis, M. K. Abekoe, and K. E. Giller, "Evaluating sustainable and profitable cropping sequences with cassava and four legume crops: effects on soil fertility and maize yields in the forest/savannah transitional agro-ecological zone of Ghana," Field Crops Research, vol. 103, no. 2, pp. 87–97, 2007.

10. D. Hounkonnou, D. K. Kossou, T. W. Kuyper et al., "Convergence of sciences: the management of agricultural research for small-scale farmers in Benin and Ghana," NJAS-Wageningen Journal of Life Sciences, vol. 53, no. 3-4, pp. 343–367, 2006.

11. P. Mapfumo, F. Matambanengwe, R. Chikowo et al., "Farmers' perceptions lead to experimentation?"LEISA Magazine, vol. 24, 2008.

12. S. Adjei-Nsiah, C. Leeuwis, O. Sakyi-Dawson, K. E. Giller, and T. W. Kuyper, "Exploring diversity among farmers for orienting inter-

disciplinary action research on cropping system management in Wenchi, Ghana: the significance of time horizons," International Journal of Agricultural Sustainability, vol. 5, pp. 176–194, 2007.

13. S. Adjei-Nsiah, "Evaluating cropping sequences with cassava and three grain legume crops: effects on soil fertility and maize yields in the semi-deciduous forest zone of Ghana," Journal of Soil Science And Environmental Management, vol. 4, no. 1, 2012.

14. S. Adjei-Nsiah, Cropping systems, land tenure and social diversity in Wenchi, Ghana: implication for soil fertility management [Ph.D. thesis], Wageningen University, The Netherlands, 2006.

15. A. A. Yusuf, R. C. Abaidoo, E. N. O. Iwuafor, O. O. Olufajo, and N. Sanginga, "Rotation effects of grain legumes and fallow on maize yield, microbial biomass and chemical properties of an Alfisol in the Nigerian savanna," Agriculture, Ecosystems and Environment, vol. 129, no. 1–3, pp. 325–331, 2009.

16. S. Recous, D. Robin, D. Darwis, and B. Mary, "Soil inorganic N availability: effect on maize residue decomposition," Soil Biology and Biochemistry, vol. 27, no. 12, pp. 1529–1538, 1995.

17. W. D. Sakala, G. Cadisch, and K. E. Giller, "Interactions between residues of maize and pigeonpea and mineral N fertilizers during decomposition and N mineralization," Soil Biology and Biochemistry, vol. 32, no. 5, pp. 679–688, 2000.

18. K. E. Giller and G. Cadisch, "Future benefits from biological nitrogen fixation: an ecological approach to agriculture," Plant and Soil, vol. 174, no. 1-2, pp. 255–277, 1995.

19. N. Ae, J. Arihara, K. Okada, T. Yoshihara, and C. Johansen, "Phosphorus uptake by pigeon pea and its role in cropping systems of the Indian subcontinent," Science, vol. 248, no. 4954, pp. 477–480, 1990.

20. P. Mapfumo, B. M. Campbell, S. Mpepereki, and P. Mafongoya, "Legumes in soil fertility management: the case of pigeonpea in smallholder farming systems of Zimbabwe?" African Crop Science Journal, vol. 9, no. 4, pp. 629–644, 2001.

21. C. A. Palm, K. E. Giller, P. L. Mafongoya, and M. J. Swift, "Management of organic matter in the tropics: translating theory into practice," Nutrient Cycling in Agroecosystems, vol. 61, no. 1-2, pp. 63–75, 2001.

Chapter 7

The Role of Earthworms in Tropics with Emphasis on Indian Ecosystems

Radha D. Kale[1] and Natchimuthu Karmegam[2]

[1]Centre for Scientific Research and Advanced Learning, Mount Carmel College, Bangalore, Karnataka 560 052, India

[2]Department of Biotechnology, VMKV Engineering College, Vinayaka Missions University, Periya Seeragapadi, Salem, Tamil Nadu 636 308, India

ABSTRACT

The paper highlights the research carried out by different scientists in India on aspects of earthworm population dynamics and species diversity, associated with other soil fauna and microflora. It also deals with the importance of earthworm activity on physicochemical properties of soil with reference to India and other tropical countries. Stress is laid on the earthworm plant association and importance of the secretions of earthworms as plant growth stimulators. Moreover, the earthworm species reported and being utilized for vermicomposting in India are discussed, since vermicomposting is the ultimate technology which renders for the improvement of soil fertility status and plant growth. Earthworms serve as indicators of soil status such as the level of contamination of pollutants: agrochemicals, heavy metals, toxic substances, and industrial effluents; human-induced activities: land-management practices and forest degradation. In all these fields there is lacuna with respect to contributions from India when compared to the available information from other tropical countries. There is lot of scope in the field of research on earthworms to unravel the importance of these major soil macrofauna from holistic ecological studies to the molecular level.

1. INTRODUCTION

Earthworms belonging to Phylum Annelida, Class Chaetopoda, and Order Oligochaeta occupy a unique position in animal kingdom. They are the first group of multicellular, eucoelomate invertebrates who have succeeded to inhabit terrestrial environment. They form major soil macrofauna. Their species richness, abundance, and distribution pattern reflect on edaphic and climatic factors of the geographical zone. They serve as “bioindicators” to understand the physicochemical characteristics of their habitat. Their horizontal and vertical stratification and abundance contribute to pedogenesis and soil profile. Encouraging their establishment through no tillage or shallow ploughing and enriching soil with organic matter incorporation has resulted in improving soil fertility. This has been experimented for several decades at Rothamsted Research Station, U.K. The interaction of earthworms and other microflora and fauna has given much scope for understanding of soil community and its influence on above ground primary production.

Distinctive habitat, food niches, and adaptive mechanisms of earthworms have opened up new fields for investigations on their role in organic waste management. One of the advantageous factors in this field is the use of earthworms to minimize the degradable organic matter and to use the same as bioresource for organic manure production. The manure produced serves as good source of soil amendment. The ecologically distinguished epigeic earthworms are used for producing the organic manure, “vermicompost”. This has gained attention of garden lovers, agriculturists, and agroindustries to convert organic matter generated at different levels into rich, odorless, free flowing compost to support sustainable agriculture.

2. EARTHWORMS: COMPONENTS OF SOIL BIOTA

Earthworms form one of the major macrofauna among soil biota to maintain dynamic equilibrium and regulate soil fertility. Their existence depends on adequate moisture, soil texture, pH, electrolyte concentration, and food source in the given ecosystem. This clearly indicates the interdependency of the environmental factors to the survival of earthworms; when such conditions are created, they further contribute to soil fertility through their activity.

3. FOOD NICHES OF EARTHWORMS

Degradation of leaf material commences from the time it detaches itself from the plant and drops to ground to add to litter. Earthworms are the major secondary decomposers in the soil faunal community. They feed on decomposed organic material at different levels of degradation. Lee [1] has suggested that earthworms survive on microorganisms, micro- and mesofauna associated with ingested dead tissue. According to him, earthworms that feed near the surface on decomposing litter and at the root zone on dead roots are the detritivores and those remain at subsurface and consume large quantities of soil are geophagous earthworms.

Lavelle [2] has categorized geophagous earthworms as polyhumic, oligohumic, and mesohumic based on the proportion of humus and soil in their feed. Through factorial analysis, he has given the explanation that temperature differences with latitude and litter characteristics like quantity and decomposability determine the variations observed with reference to their distribution. The detritivorous epigeic earthworms form the major component of earthworm fauna in temperate regions and mesohumic endogeic earthworms are predominant in tropical forests. There is minimum representation of mesohumic earthworms in temperate regions. Oligohumic earthworms that feed on soil having very low level of organic matter are abundant only in tropical regions.

Lavelle [2] considers polyhumic earthworms as more stable fraction of earthworm community occupying different soil strata as topsoil feeders to species of rhizosphere in tropical regions. Thus, tropical earthworms depend more on soil mixed with different levels of humic substances rather than surface litter. More stable environments like heavy rainfall areas (2000 to above 4000 mm rain/annum) in the state of Karnataka, India, have greater diversity of earthworms than the dry areas (<600 to 900 mm rain/annum). The geophagous earthworms of mesohumic and polyhumic types are widely distributed in places receiving heavy rainfall in this subtropical part of the country (Tables 1 and 2).

The acceptance level of various leaf litters shows positive correlation to nitrogen and carbohydrate contents and negative correlation to polyphenol content [3]. Ganihar [4] studied the litter feeding of Pontoscolex corethrurusin a multiple-choice test. He found variations in degree of acceptability of different litter that showed positive correlations to levels of organic carbon and nitrogen content. The least preference for Eucalyptus camaldulensisand Acacia auriculiformis was linked with high levels of polyphenols. It has been shown that Lampito mauritiiexhibited similar preference either for partially decomposed large pieces of leaf material of different types or for powdered leaves mixed with agar base [5]. It could be inferred that apart from physical nature of leaf matter, chemical compounds in them serve as attractants or repellants (Tables 3 and 4). Ganihar [4] is of the view that in land reclamation sites, if earthworms have to be introduced, it is essential to develop above ground plant community. Litter from such plants when mixes with soil, at different levels of decomposition, serves as feed to developing earthworm population. The available carbon source encourages population growth of earthworms [6]. In India, Lampito mauritii is the most widely distributed earthworm in different agro-ecosystems [7–12]. This earthworm preferred decomposing grass of paddy (Oryza sativa) and finger millet (Eleucine coracana) to other leaf litter [5]. The grasses when developed in reclamation sites can form an ideal base for establishment ofLampito mauritii to bring about improvement in soil structure and finally chemical and biological activities. Food preference and sensitivity to other edaphic factors determine the possibility of introduction of earthworms for land reclamation.

Table 1. Earthworm distribution in Southern Karnataka (India) in different agroclimatic zones including coastal plains, hilly regions, and interior plains.

Sl. No.	Species	Moisture level (%)	Soil type	Vertical distribution (cm)	Food niche	Population density (no./100 m^2)
1	*Curgeona narayani*	Wet land-in waterlogged soil	Red loamy soil	Up to 45	Mesohumic	640–11,250
2	*Dichogaster affinis*	20–40	Red loamy, alluvial and lateritic	5–10	Mesohumic to polyhumic	60–250
3	*D. bolaui*	20–40	″	″	″	60–450
4	*D. curgensis*	20–40	Red loamy	″	Polyhumic	25–200
5	*D. modigliani*	20–40	Red sandy	″	Mesohumic	10–25
6	*D. saliens*	20–40	Red sandy	″	″	65–265
7	*Drawida ampullacea*	>40	Red loamy	10–20	Polyhumic	275–930
8	*D. barwelli*	>50	Red loamy to sandy soil	10–30	″	275–576
9	*D. barwelli impertusa*	>50	Red loamy	″	″	120–430
10	*D. calebi*	>50	Red loamy to sandy soil	10–30	Polyhumic	80–1200
11	*D. ferina*	40–50	Red loamy	20–30	Mesohumic	40–340
12	*D. ghatensis*	40–50	″	10–20	″	450–1350
13	*D. kanarensis*	40–50	″	″	″	85–400
14	*D. lennora*	40–50	Red sandy soil	″	″	15–30
15	*D. modesta*	40–50	″	10–30	″	4–500
16	*D. paradoxa*	>40	Red loamy to alluvial	10–20	Polyhumic	1700–2500
17	*D. pellucida pallida*	>40	Lateritic to Red loamy	″	Mesohumic	4–500
18	*D. scandens*	>40	Red sandy loam	5–10	Polyhumic	10–350
19	*D. sulcata*	>40	Alluvial soil	10–30	Polyhumic	65–235

20	*Glyphidrillus annandalei*	>40	Sandy bed to Red loam	20–45	Oligohumic	130–1600
21	*Gordiodrilus elegans*	>40	Red sandy loam	10–40	Mesohumic	24–200
22	*Hoplochaetella kempi*	30–40	Lateritic to alluvial	10–30	Polyhumic	10–430
23	*H. suctoria*	30–40	Alluvial	10–20	″	50–240
24	*Hoplochaetella sp.*	40–50	Red loam	20–40	″	460–3330
25	*Howascolex sp.*	30–40	Red loam	10–30	″	145–2500
26	*Lampito mauritii*	20–30	Red sandy to lateritic	10–30	Mesohumic	720–2190
27	*Mallehula indica*	30–40	Red loam	10–20	Mesohumic	180–880
28	*Megascolex filiciseta*	30–40	Lateritic	5–10	Polyhumic	15–330
29	*M. insignis*	30–40	Alluvial	5–20	Polyhumic	65–800
30	*M. lawsoni*	30–40	Red loam to sandy loam	10–30	Mesohumic	120–1000
31	*M. konkanensis*	30–40	Lateritic to alluvial	20–45	Mesohumic	20–3900
32	*Metaphire houlleti*	>40	Alluvial and Red loam	10–40	Polyhumic	18–2140
33	*Octochaetona albida*	30–40	Red loam	10–20	Polyhumic	150–650
34	*O. beatrix*	20–30	Sandy loam	″	″	40–335
35	*O. rosea*	30–40	Alluvial	10–20	Mesohumic	15–120
36	*P. excavatus*	>40	Organic layer	0–5	Detritivore	18–8000
37	*Plutellus timidus*	30–40	Alluvial	10–15	Mesohumic	60–460
38	*Polypheretima elongata*	>40	Sandy loam to Red loam	30–60	Mesohumic	194–4000
39	*Pontoscolex corethrurus*	30–50	Sandy, alluvial, loamy, lateritic	5–15	Mesohumic to polyhumic	250–7100

Table 2. Habitat preference of widely distributed earthworm species Lampito mauritii andPontoscolex corethrurus at study sites.

District	Agroclimatic zone	Mean annual rainfall (mm)	Soil type	Earthworm species	Habitat preference
Bangalore	Eastern dry zone	700–900	Red loamy soil	*Lampito mauritii*	Arable lands
Kolar	"	600–800	Lateritic and red sandy soil	"	Grasslands
Tumkur	"	"	Red sandy soil	"	Grasslands and arable lands
Chickmagalur	South transition zone	900–1000	Red loamy soil	Species richness than species dominance	Varied habitats
Chickmagalur	Hilly zone	2000–3000	Red loamy soil	*Pontoscolex corethrurus*	Plantations
Coorg	Hilly zone	2000–>4000	"	"	Grasslands and plantations
South Kanara	Coastal zone	3000 > 4000	Coastal alluvial soil	*P. corethrurus* and *Megascolex konkanensis*	Grasslands, plantations, arable lands

Table 3. Disintegration of different leaf matters due to selective feeding by earthwormLampito mauritii [5].

Leaf matter	1	2	3	4	5	6	7	8	9
Millet straw	70.00	50.00	55.00	—	—	—	—	—	—
Paddy straw	48.00	11.00	27.50	22.00	13.00	33.00	—	—	—
Cashew litter	—	—	—	38.00	24.00	39.00	22.50	2.60	67.00
Mango litter	—	—	—	48.00	30.00	50.00	30.70	28.60	6.30
Guava litter	—	—	—	44.00	14.00	83.00	25.00	23.00	—
Eucalyptus litter	—	—	—	32.00	10.00	61.00	31.00	24.40	11.60

Note: *Col: 1–3 data for first month*

(1) Percent loss of litter per month due to microbial degradation and feeding by earthworms.
(2) Percent microbial degradation per month.
(3) Rate of litter consumption (mg) for hundred earthworms per day.

Col: 4–6 data for second month

(4) Percent loss of litter per month due to microbial degradation and feeding by earthworms.
(5) Percent microbial degradation per month.
(6) Rate of litter consumption (mg) for hundred earthworms per day.

Col: 7–9 data for third month

(7) Percent loss of litter per month due to microbial degradation and feeding by earthworms.
(8) Percent microbial degradation per month.
(9) Rate of litter consumption (mg) for hundred earthworms per day.

The table also shows the acceleration of litter breakdown in presence of earthworms.

Table 4. Artificial diet (1: 8 by weight) of agar and different leaf litter powder on feeding of earthworm Lampito mauritii in relation to C/N of diets [5].

Litter powder in agar	Daily food intake mg/day/adult	C/N of the feed
Paddy straw	8.05 ± 0.28	37
Millet straw	7.07 ± 1.23	45
Mango litter	8.67 ± 1.27	19
Guava litter	3.25 ± 0.79	45
Cashew litter	4.44 ± 1.10	30
Eucalyptus litter	1.62 ± 0.59	42
Agar only (control)	2.53 ± 1.23	38

Number of observations = 3; Palatability depends on texture as well as chemical nature of the feed.

4. EARTHWORM ACTIVITY ON PHYSICOCHEMICAL PROPERTIES OF SOIL

Earthworms are the major macrofauna in the soil community. They are distributed at different depths in soil strata. The litter feeders, which are not burrowers, constitute a very small number in tropical situations. The burrowing endogeic earthworms live in horizontal and vertical burrows constructed in soil strata. They make these burrows partly by ingesting soil particles through their way and partly by pushing the soil to the sides [13]. The ingested soil along with organic matter passes through the gut and undigested matter is released at the opening of the burrow on soil surface or at the subsurface as castings. The subsurface castings contribute to soil profile [1].

The burrows of earthworms, which run horizontally or vertically depending on burrow forming ability of species, will determine the possible physical effects on soil characteristics. In temperate regions where deep burrowing anecic earthworms are of common occurrence, it is opined that infiltrations can bring about leaching of nutrients from soils to ground water. The leachate volume may show an increase of four to twelve folds due to their activity [14]. Introduction of Aporrectodea caliginosa into coniferous forest soils resulted in fifty fold increase in concentration of nitrate and cations in soil solution. But the amount that entered ground water or plant system remained undetermined [15]. One of the major contributions of burrowing activity of earthworms is in affecting soil porosity [16, 17]. The major impact on hydrology has been worked out with respect to activity of anecic earthworm Lumbricus terrestris [18]. Information is lacking in India with respect to burrows of earthworms, their structure, and any variations observed depending on soil type. Influence of organic matter, agricultural practices on earthworm population, and similarly the role of earthworms in modifying the situations in cultivable lands are very meager in a country having diversity and abundance of the populations in different agro-ecosystems. Reddy et al. [19] reported the influence of various management practices affecting density and surface cast production. The casts of the earthworm, Pontoscolex corethrurus, and the surrounding soil in an undisturbed forestfloor in Sirumalai Hills, Tamil Nadu (South India) showed that the percentage of moisture content, organic carbon, and total nitrogen in the worm casts were higher and significantly differed from the values obtained in the surrounding soil [20].

According to the recent report by Julka et al. [21], in India, there are 590 species of earthworms with different ecological preferences, but the functional role of the majority of the species and their influence on the habitat are lacking. Recently Karmegam and Daniel [11] reported the correlation of soil and environmental parameters on the abundance of ten different earthworm species belonging to four families, namely, Megascolecidae (Lampito mauritii, L. kumiliensis, and Megascolex insignis), Octochaetidae (Dichogaster bolaui, D. saliens, andOctochaetona thurstoni), Moniligastridae (Drawida chlorina, D. paradoxa, and D. pellucida pallida), and Glossoscolecidae (Pontoscolex corethrurus) in the study that was carried out at different locations in Dindigul District (South India). The fluctuations in populations of earthworms were observed during the monthly collections in course of three years in all the selected sites. In the survey carried out from 1997 to 1999, the predominant species that were recorded as maximum number of earthworms/m2 in sites 1–10 were D. pellucida pallida (Jan. 1998-70.44), D. pellucida pallida (Dec. 1999-32.30), L. mauritii (Feb. 1998-55.22), D. pellucida pallida (Dec. 1999-25.54), L. mauritii (Dec. 1997-66.78), L. mauritii (Nov. 1997-43.40), L. mauritii (Jan. 1999-44.60), P. corethrurus (Nov. 1997-58.34), P. corethrurus (Dec. 1999-64.30), and P. corethrurus (Dec. 1998-107.60) [22].

The biomass dynamics also showed wide fluctuation among the species in relation to the months of collection from different collection sites. The highest worm biomass was recorded during December to February and certain species were totally absent during certain periods of the survey. The total biomass of different species recorded in the monthly observation over a period of three years (1997 to 1999) varied in various study sites. The highest biomass of the respective earthworm species as well as the month and year of its occurrence in the study sites 1 to 10 as recorded includes D. pellucida pallida (30.63 g/m2 during Feb. 1998), D. pellucida pallida(22.88 g/m2 during Jan. 1998), D. pellucida pallida (29.27 g/m2 during Dec. 1999), D. pellucida pallida (20.20 g/m2 during Dec. 1999), D. pellucida pallida (44.65 g/m2 during Dec. 1999), D. pellucida pallida (22.38 g/m2during Dec. 1999), D. pellucida pallida (29.66 g/m2 during Jan. 1998), P. corethrurus (15.20 g/m2 during Dec. 1998), D. bolaui (19.79 g/m2 during Jan. 1999), and P. corethrurus (26.34 g/m2 during Dec. 1998), respectively [22]. Among the earthworm species studied, L. kumiliensis has been reported for the first time in Sirumalai Hills of Tamil Nadu, India [23]. This is the only study to highlight the cyclic fluctuations in the earthworm populations for a continuous period of three years and variations in the species structure at different time intervals. Still the information on the physicochemical changes in the soil with respect to species composition at given time is not clear. A composite study on microbial association with the predominant earthworm species at a given time may provide necessary information on its ecological role.

5. FACTORS INFLUENCING THE ABUNDANCE OF EARTHWORM POPULATIONS

The percentage abundance of different species of earthworms in the 10 collection sites during the survey period (1997–1999) is shown in Figures 1 and 2. In mostof the study sites, that is, 1–7, L. mauritii was the dominant species and it showed its presence during the premonsoon, monsoon, and postmonsoon months. P. corethrurusshowed its abundance in the sites 8–10. Various parameters, that is, pH, electrical conductivity (EC), organic carbon (OC), nitrogen (N), atmospheric temperature (AT), soil temperature (ST), soil moisture (SM), humidity (HUM), and rainfall (RF) observed during the survey period (1997–1999) are given in Table 5 and in Figure 3. All the parameters showed fluctuations in all the ten study sites. Here, for the convenience of statistical analysis the parameters were categorized into two major groups: (a) physicochemical parameters which included pH, EC, OC, and N; and (b) climatic parameters which included ST, SM, HUM, and RF.

Table 5. Physicochemical and climatic characteristics (average) of the study sites 1 to 10 (1997–1999) (refer to Table 6 for study site description) [22].

Parameter observed*	Study sites									
	1	2	3	4	5	6	7	8	9	10
1997										
pH	7.78	7.63	7.13	6.86	7.59	7.67	6.78	7.04	7.50	6.55
EC (dS/m)	0.34	0.20	0.38	0.11	0.21	0.18	0.32	0.14	0.27	0.39
OC (%)	1.42	2.29	4.44	2.75	1.47	2.94	3.42	3.05	4.20	7.99
TN (%)	0.41	0.35	0.23	0.23	0.25	0.40	0.42	0.31	0.22	0.40
ST (°C)	28.90	29.83	27.84	29.29	30.31	29.27	29.83	23.47	22.49	21.30
SM (%)	8.10	6.34	10.34	7.22	7.16	8.30	10.25	15.75	15.46	14.99
1998										
pH	7.95	7.51	7.25	6.66	7.51	7.62	6.45	7.15	7.34	6.44
EC (dS/m)	0.36	0.22	0.38	0.13	0.25	0.16	0.33	0.18	0.31	0.39
OC (%)	1.74	2.19	4.24	2.79	1.43	2.35	4.25	3.19	4.22	8.48
TN (%)	0.44	0.34	0.24	0.27	0.27	0.41	0.41	0.27	0.27	0.38
ST (°C)	29.23	30.18	28.27	30.63	29.92	29.30	30.18	24.14	22.68	21.30
SM (%)	12.14	9.93	15.18	9.73	12.83	12.03	14.35	16.00	17.09	16.29
1999										
pH	7.85	7.49	7.37	6.85	7.38	7.59	6.47	6.98	7.45	6.64
EC (dS/m)	0.35	0.23	0.34	0.14	0.24	0.17	0.35	0.16	0.26	0.39
OC (%)	1.40	2.45	4.34	2.90	1.36	3.02	4.05	3.45	4.37	9.99
TN (%)	0.46	0.37	0.27	0.26	0.26	0.43	0.41	0.31	0.28	0.39
ST (°C)	27.42	28.55	26.49	29.66	30.42	27.46	28.55	25.21	23.51	22.80
SM (%)	9.50	7.27	11.70	8.50	9.27	9.37	10.34	12.56	14.37	13.86

* EC: Electrical conductivity; OC: Organic carbon; TN: total nitrogen; ST: Soil temperature; SM: Soil moisture.

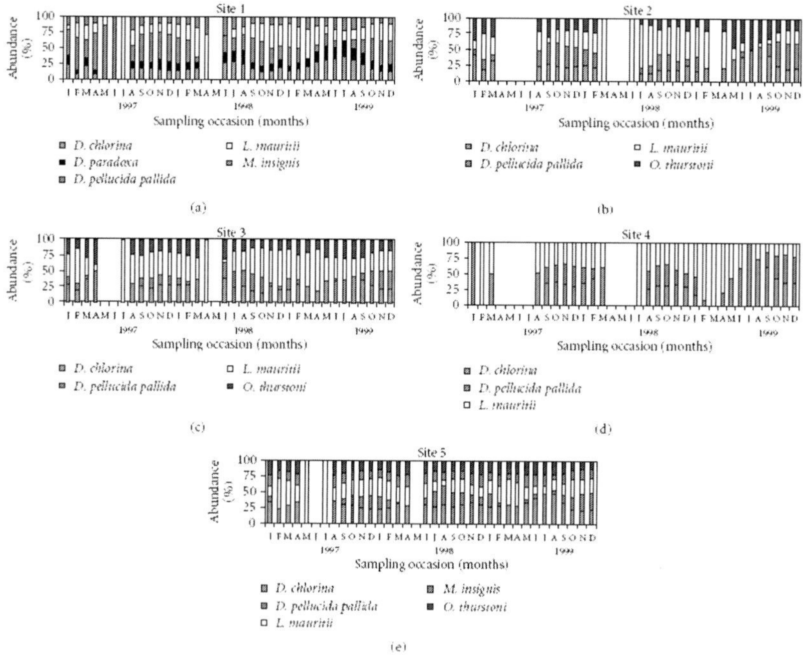

Figure 1. Percentage abundance of earthworm population in study sites 1 to 5 (1997–1999).

In Tamil Nadu, India, very limited information is available on the distribution pattern of earthworms. The data on earthworm distribution is available for the stations like Palni Hills [24], Madras [25], and Sirumalai Hills [11, 23, 26, 27]. Dindigul, a District in Tamil Nadu, was considered as study site for its variety of habitats to assess the earthworm species diversity, density, and biomass. The population and biomass dynamics of different earthworm species and their percentage abundance in relation to physicochemical characteristics of the soil and the climatic factors were recorded in selected sites. The correlation of earthworm population to physicochemical characteristics of the soil and the climatic parameters was carried out to find out the possibility of arriving at a suitable endemic earthworm species for vermicomposting operations in this part of the country. Since the populations of earthworms are extremely variable in size ranging from only a few individuals (sometimes totally absent) to more than 1000 /m2, the assessment of the size distribution and structure of earthworm population is difficult. The seasonal change, demography, and vertical distribution of the populations make it more complicated, and hence, it is absolutely essential to follow a uniform method of determining the number of earthworms in small sample areas as it has been done in this study. The regular monthly survey carried out for three years (1997 to 1999) showed the presence of ten species of earthworms, with four species restricted only to the hilly region and six species to the plain, including the foothills (Table 6). This observation indicates that species such as L. kumiliensis, D. bolaui, D. saliens, and P. corethrurus are specific only to the hilly region and they are not found in the foothills. Though L. kumiliensis and L. mauritii both belong to the same genus, Lampito, L. kumiliensis was found only in the hilly region and L. mauritii in the plains. This observation indicates that the distribution of different earthworm species is limited even though they are closely related. Such niche differences for closely related species have been reported by earlier workers in the field [28, 29].

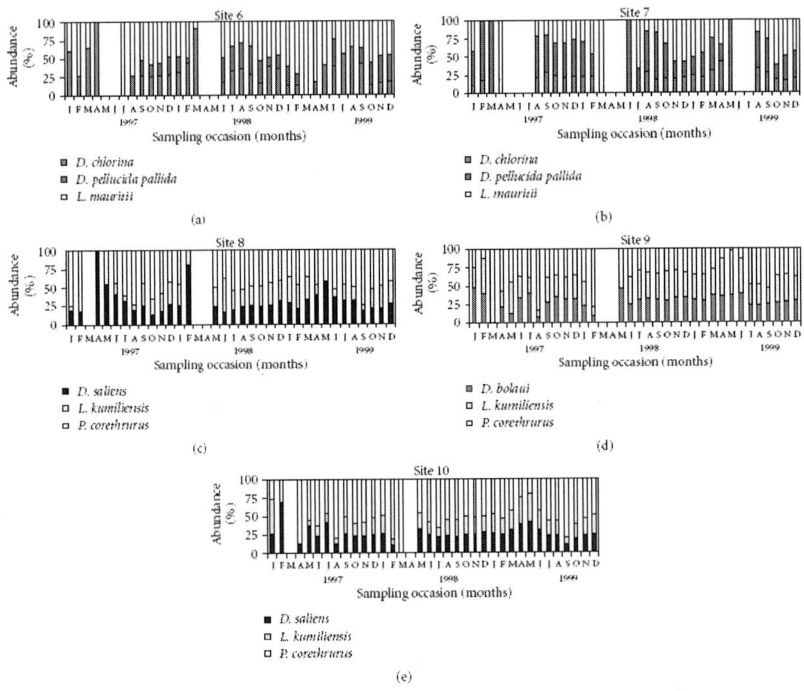

Figure 2. Percentage abundance of earthworm population in study sites 6 to 10 (1997–1999).

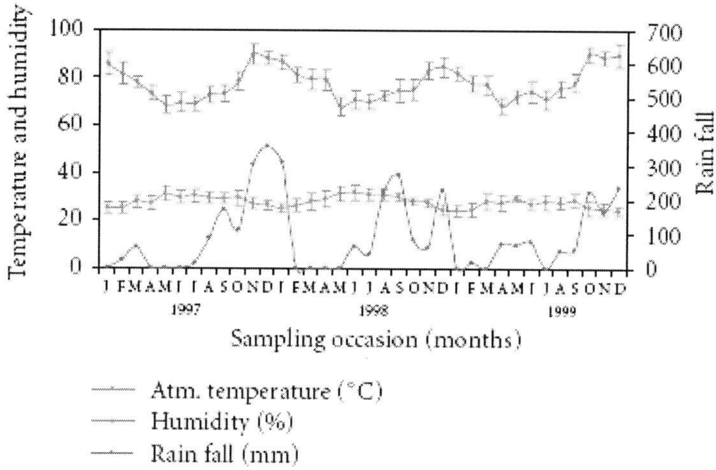

(a)

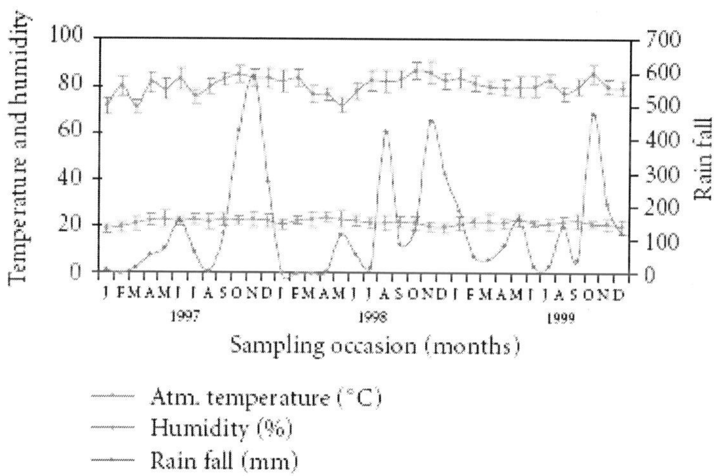

(b)

Figure 3. Atmospheric temperature (mean ± SD), Humidity (mean ± SD), and average rainfall of the study sites 1–7 (a) and 8–10 (b).

The results of the percentage abundance of different species of earthworms showed that L. mauritii and P. corethrurus were the most abundant in the study sites 1 to 7 and 8 to 10, respectively. Formation of aggregation of species has been observed in sites 1 to 7; that is, wherever L.

mauritii was found, it was in association with D. chlorina and D. pellucida pallida. This sort of association of earthworm species sharing the same habitat is not uncommon [1, 30]. L. mauritii is the dominant species found almost all over India along with other earthworm species such as Drawida modesta, Octochaetona pattoni, O. thurstoni, Ramiella pachpaharensis, Polypheretima elongata, and Pontoscolex corethrurus [8, 31], but Bano and Kale [32] reported that L. mauritii was not found in some forest areas and coastal Karnataka. The population densities of earthworms observed in the 10 collection sites ranged from 0 to 228 /m2. Other authors observed population densities (earthworm no./m2) of 53.5 in plain grass land, 73 in deciduous forest, 543 in the fallow phases of shifting agriculture, and 58.2 in the maize crop land [33–36]. In rubber plantations of Tripura (India) about 20 species of earthworms, namely, Eutyphoeus gigas, E. gammiei, E. comillahnus, E. assamensis, E. festivas, Eutyphoeus sp., Dichogaster bolaui, D. affinis, Lennogaster chittagongensis, Octochaetona beatrix, Metaphire houlleti, Perionyx sp., Kanchuria sumerianus, Kanchuria sp.1, Kanchuria sp.2, Drawida nepalensis, Drawida sp.1, Drawida sp.2, Pontoscolex corethrurus, andGordiodrilus elegans were distributed and it was observed that the largely dominating species were endogeics [37].

Evans and Guild [38] have shown that nitrogen rich diets help in rapid growth of earthworms and facilitate more cocoon production than those with little nitrogen available. Due to the influence of nitrogen content of the soil, the percentage contribution of nitrogen to earthworm population might have shown a very high degree of dependence in the present study. Some of the reports from the country well support qualitative dependence of earthworm population on soil nitrogen content [26, 27, 39, 40].

Soil moisture plays a major role in the distribution and occurrence of various earthworm species. The same has been observed by other workers in their studies [25, 28, 29, 41, 42]. The abundance and species diversity are dependent on climatic conditions, especially the occurrence of dry and/or cold periods, and regional variation in vegetation, soil texture, and nutrient content. The climatic parameters, that is, soil temperature, soil moisture, humidity, and rainfall show seasonal fluctuations (Table 6 and Figure 3). The highest rainfall was recorded during October-November and the earthworm population was also the highest at this period. The soil moisture content corresponded with earthworm population. Total annual rainfall of 1130, 1284, and 959 mm was recorded during 1997, 1998, and 1999 in the plains and foothills of Sirumalai (study sites 1–7). The highest rainfall of 304 and 357 mm was received during October and November 1997 in the above study sites. The highest rainfall months in Sirumalai Hills (study sites 8–10) were October to December. The soil moisture content directly matched with the rainfall. The soil moisture content ranged from 2.0 to 30.4 percent in the study sites 7–10 during the three years of the study. The humidity also showed fluctuations in both the plains and hilly region of the study area. Soil moisture can explain the increase in earthworm population, since soils are moist under a mulch cover because of the restricted evaporation. There are many indications, to show that the population of endogeic earthworms is controlled mainly by soil moisture [42].

The influence of climatic factors on the populations of earthworm is not uncommon. The populations ofMillsonia anomala are dependent on climatic conditions as well as vegetational patterns. Earthworm activity and populations are determined essentially by the moisture content of the soil [43]. The temperature and moisture are usually inversely related and higher surface temperature and dry soils are limiting factors to earthworms than low and water logged soils [44]. The soil temperature plays an important role in the maintenance of earthworm population in an ecosystem and available information also indicates the negative correlation of soil temperature to earthworm population [11, 25, 40, 45]. In rubber plantations of Tripura (India), the earthworms experienced 2 5 . 9∘C, 24.8%, 4.85, and 1.8% mean soil temperature, moisture, pH, and organic matter, respectively [37]. Temperature largely affects activity of earthworms in temperate regions. Tropical species can withstand higher temperatures. L. mauritii is available throughout the year where the annual temperature is 3 0 ± 2∘C. Population of O. serrata was active between 27 and 2 8∘C. In tropical regions the temperature fluctuations are minimal when compared to temperate regions.

Moisture is another limiting factor for earthworm distribution as water constitutes a major portion of the body weight of an earthworm. Soil moisture and population estimates are positively correlated [35]. Water constitutes 75–90 percent of the body weight of earthworms. So the prevention of water loss is a major factor for their survival. They apparently lack a mechanism to maintain constant internal water content, so that their water content is influenced greatly by the water potential of the soil [46], which directly depends on the adequate availability of soil moisture.

Table 6. Population density of earthworms in different habitats in Dindigul District, Tamil Nadu studied during 1997–1999 [22].

Study site	Description	Earthworm species	Avg. population density (no./m^2)
(1)	Cultivated land	*Lampito mauritii* (Kinb.).	12.52
		Megascolex insignis Mich.	7.82
		Drawida chlorina (Bourne).	8.88
		Drawida paradoxa Rao.	5.10
		Drawida pellucida var. *pallida* Mich.	18.60
(2)	Unirrigated crop land	*Lampito mauritii* (Kinb.).	14.18
		Octochaetona thurstoni Mich.	5.46
		Drawida chlorina (Bourne).	5.04
		Drawida pellucida var. *pallida* Mich.	11.10
(3)	Uncultivated shaded fallow land	*Lampito mauritii* (Kinb.).	17.88
		Drawida pellucida var. *pallida* Mich.	13.27
		Octochaetona thurstoni Mich.	10.92
		Drawida chlorina (Bourne).	10.70
(4)	Uncultivated fallow land	*Lampito mauritii* (Kinb.).	10.30
		Drawida chlorina (Bourne).	4.73
		Drawida pellucida var. *pallida* Mich.	6.46
(5)	Garden	*Lampito mauritii* (Kinb.).	15.50
		Megascolex insignis Mich.	10.96
		Octochaetona thurstoni Mich.	13.04
		Drawida chlorina (Bourne).	17.26
		Drawida pellucida var. *pallida* Mich.	11.27

(6)	Orchard	*Lampito mauritii* (Kinb.).	8.92
		Drawida chlorina (Bourne).	6.32
		Drawida pellucida var. *pallida* Mich.	4.75
(7)	Foothills (Alt. < 450 m.)	*Lampito mauritii* (Kinb.).	5.63
		Drawida chlorina (Bourne).	6.22
		Drawida pellucida var. *pallida* Mich.	22.68
(8)	Grass land (Alt. 1,000 m.)	*Lampito kumiliensis* (Kinb.).	18.21
		Dichogaster saliens (Bedd.).	5.31
		Pontoscolex corethrurus (Muller).	10.30
(9)	Semi-evergreen forest (Alt. 1,100 m.)	*Lampito kumiliensis* (Kinb.).	29.52
		Pontoscolex corethrurus (Muller).	10.49
		Dichogaster bolaui (Mich.).	9.39
(10)	Sacred grove land (Alt. 1,300 m.)	*Lampito kumiliensis* (Kinb.).	19.42
		Dichogaster saliens (Bedd.).	9.37
		Pontoscolex corethrurus (Muller).	19.16

The seasonal dynamics in an annual cycle shows that earthworm numbers and biomass were high in the rainy season with a gradual decline in number in the winter season. Earthworms were completely absent during the second half of January and February, when soil temperature was very low (4.9–6 . 2oC). Dash and Patra [7] and Kale and Krishnamoorthy [8, 47] have recorded maximum number of earthworms and biomass in the rainy and late rainy period. The relationship between earthworm activity and rainfall was observed by Fragoso and Lavelle [48] and Joshi and Aga [49]. The moisture requirements for different species of earthworms from different regions can be quite different [42]. The dependence of earthworm population on soil moisture is seen in the studies carried out for three years as of the highest degree when compared with other climatic parameters. This is because of certain physiological activities of earthworms such as cutaneous respiration and excretion of nitrogenous ammonia and urea, which need a moist environment, which, in turn, is essential for the maintenance of their life process.

Systematic correlation analysis results indicate that only about 80 percent of the population dependence can be explained by these physicochemical and climatic parameters and it is presumed that the remaining may depend on other environmental factors. The correlation analysis technique may be used to quantify and rationalize the effects of physicochemical parameters on the earthworm population. However, no single factor is likely to be solely responsible for the horizontal distribution of earthworms, but rather the interaction of several of the factors provides suitable soil conditions for the existence of earthworm populations [11].

6. EARTHWORM CASTS: ABUNDANCE, STRUCTURE, AND PROPERTIES

Earthworms' release "cast" at the opening of their burrows. Epigeic earthworms release the castings exclusively on soil surface. Their castings may be granular or spindle like masses that may be 2 to 3 cm high heaps as inEudrilus eugeniae or Perionyx excavatus. There is no definite shape to the excreted matter to identify as castings of Eisenia fetida. Eisenia fetida releases fine, powdery, dark brown material as surface cast. Soil living endogeic earthworms that feed on different quantities of organic matter along with soil particles use part of their castings to strengthen their burrow walls and the rest is released as castings. Castings of these earthworms may be ovoid or irregularly shaped minute mounds. Though the nature of cast released is characteristic of a species, this cannot be criterion for their identification [50]. If pellet-like castings are released by Pheretima posthuma, Perionyx millardi releases thread-like castings. Thick and long winding columns of hollow mound of 5 cm long and 2.5 cm wide casts are characteristic of Hoplochaetella khandalaensis. The biggest cast of Notoscolex birmanicus weighing 1.6 kg after drying for four months is reported from Burma [50]. Polypheretima elongataand Pontoscolex corethrurus excrete the ingested soil as sticky, thick lumps on soil surface.

Amount of cast produced can serve as an index for assessing earthworm activity. Immediately after rains, release of surface casts will be at a maximum level. At this point of time, majority of earthworms are found at 0 to 10 cm depth and very few of them are found at 20 to 30 cm depth (Kale and Dinesh, 2005, unpublished). Surface cast production has been quantified in different agro-ecosystems to relate it to their abundance [51–53]. Influence of seasonal variation and land use pattern was observed with respect to cast production in shifting agriculture [34]. Norgrove and Hauser [54] have recorded around 30 to 35 t/ha of cast production in tropical silvicultural system. Reddy [55] has reported annual production of 23.4 to 140.9 tonnes by Pheretima alexandri. According to Lavelle [56], cast production is rhythmic and it will be at maximum at early morning hours. In general cast production in tropical countries is restricted to wet seasons. Table 7 provides the information on earthworm cast production in different agro-ecosystems during onset of postmonsoon season in the state of Karnataka, India.

The physicochemical properties of casts depend on the habitat soil and species of earthworm [57]. Their aggregate stability depends on the available organic matter [58]. The stability of casts and stability of fragmented casts on disintegration are the important factors to determine the soil structure [1]. Aggregate stability may result from addition of mucus secretion from earthworm gut and of associated microorganisms in the gut. It may also be due to macerated organic particles in the castings that encourage microbial activity after its release from the gut [59]. According to Parle [60], stability of casts is due to fungal succession that takes place in the cast. Habibulla and Ismail [61] are of the opinion that soil texture, particle size, and porosity play an important role in burrowing and surface cast production. As casting activity is restricted to wet seasons, not much of attention is paid to assess the quantum of cast produced and its influence on soil physical, chemical, and biological properties as is available from other parts of the world. It is essential to know the physicochemical and biological variations that may be seen in cast produced by the same species of earthworm inhabiting places that differ in physiographic and edaphic conditions. This will provide the information on interrelationship of earthworms, original soil characters, and nature of available organic material that influence the change in soil characters through deposition of earthworm cast. The fertile lands turning unproductive in Himachal Pradesh, India, due to sticky castings of earthworms that turned the soil into cement-like clods had been reported [62]. Puttarudraiah and Sastry [63] had observed stunting of growth in root crops like carrot, radish, and beetroots due to castings of Pontoscolex corethrurus in pot culture studies.

Table 7. Earthworm cast production during early postmonsoon period (Nov. 2004) at different agro-ecosystems in Kuti village of Somavarpatna Taluk of Karnataka State (Kale and Dinesh 2004, unpublished).

Land uses	Castings (Kg/Sq. M)
Natural forest	11. 20 ± 0.46
Coffee plantation	17. 2 ± 0.53
Cardamom plantation	16. 80 ± 1.00
Paddy fields (after harvest)	13. 60 ± 1.00
Acacia plantation	2.40*
Grassland	0.8*

* Due to dryness prevailing at the collection spots castings could be collected only from single spots out of 6 and 8 monolith points.

Castings of earthworms are the "store house" of nutrients for plants. The increased earthworm activity with increase in availability of carbon and in turn a raise in available nitrogen and phosphorus in their castings was also reported [6]. Earthworm activity has shown to improve the soil aggregates and soil minerals that are more available to plants than from soil [54, 64]. It is clear from various studies that earthworm casts may have more important role in plant nutrition and nutrient cycling than it was assumed previously [65, 66]. In India, very early reports are available on such observations on the chemical properties of earthworm castings that can play a positive role in plant growth [57, 67, 68]. The chemical composition of casts, which is widely studied, is of holonephric lumbricid earthworms. In subtropical country like India where majority of earthworms are meronephric, their castings may show higher level of available plant nutrients than surrounding soil. Dash and Patra [7, 53] had reported higher levels of nitrogen in casts of Lampito mauritii than in surrounding soil. Ganeshmurthy et al. [69] have found higher rate of mobilization of micronutrients in earthworm castings. It requires further studies on meronephric Megascolecid earthworms and their castings on available and exchangeable forms of nutrients to assess their contribution to soil fertility. Kale and Krishnamoorthy [70] had shown increased levels of soluble calcium and carbonates in castings of Pontoscolex corethrurus. Soluble carbonates contribute to exchangeable base contents of castings (Table 8). The physicochemical properties like pH, EC, organic C, total N, available P, K, Na, Ca, and Mg of casts did not differ in zero tillage land treated with mulch of residues of annuals or perennials [19]. The population dynamics of a peregrine earthworm,Pontoscolex corethrurus, in undisturbed soil of Sirumalai Hills clearly showed that the parameters like rainfall, humidity, soil moisture, and organic carbon influence the population positively [26, 27]. It has also been reported that in rubber plantations of Tripura, a part of north-east India, Pontoscolex corethrurus was the dominant species, representing 61.5% biomass and 72% density of the total earthworm population where it might be linked to individual tree species effect (Hevea brasiliensis) that favoured P. corethrurus over other species [37].

Table 8. Calcium and carbonates in castings of Pontoscolex corethrurus compared with that of habitat soil [70].

Constituents	μg/g dry weight	
	Soil	Castings
Ionic Calcium	12.24 ± 0.41	145.50 ± 9.81
Exchangeable Calcium	12.83 ± 0.37	95.23 ± 7.28
Insoluble Calcium	179.62 ± 0.02	32.09 ± 0.93
Ionic/Insoluble Carbonate	0.15 ± 0.01	6.98 ± 2.22

7. EARTHWORMS AND MICROFLORA

Earthworm activity is closely associated with microbial activity. Lavelle [2] is of the opinion that there may exist competition between microorganisms and earthworms for easily digestible and energy rich substrates. Such competition may depend on availability of nutrients in the medium. Contrary to this, earthworms may derive benefit from microorganisms when they have to survive on materials rich in cellulose or hemi cellulose. So there exists mutualistic relation between earthworms and microorganisms. Tiunov and Scheu [6] have shown that earthworms deprive easily available carbon to microorganisms and availability of carbon increases effective mobilization of N and P by earthworms. The complex interrelationship of earthworms and microorganisms is at the level of their digestive tract, castings, and burrow walls [71]. This establishes the probable mutualism that exists between earthworms and microorganisms. Joshi and Kelkar [68] demonstrated higher microbial activity in earthworm castings and their role in mineralization of nitrogen. They incubated known weights of groundnut cake in a pot containing earthworm castings and other containing soil from the same place. The release of N from groundnut cake was at a higher level in pot containing castings than from one having soil as the medium.

Bhat et al. [72] were the pioneer contributors to report on role of microorganisms in the gut of earthworms. Khambata and Bhat [73] had made a detailed investigation on intestinal microflora of Pheretima sp. They had isolated Pseudomonas, Corenyform bacteria, Nocardia, Streptomyces, and Bacillus from the intestinal tract. There is no report of nitrosofying and nitrifying bacteria in their observations in the gut of earthworms. Dash et al. [74] have reported about isolation of 16 fungi from different parts of the gut out of 19 found in their habitat. In the fresh castings of the same earthworms there were only seven fungi with antibiotic properties or with thick spore coats. This suggests the selective fungal feeding by earthworms.

Drillosphere is the focus for understanding earthworm microbe interrelationship. This association is also associated with land use and metabolizable carbon present in the soil. Metabolizable carbon has positive effect on both microorganisms and earthworms [75]. Microbial activity will be at a higher level in the drillosphere than in surrounding soil and other edaphic factors determine the microbial diversity in drillosphere [76]. According to Kretzschmar [77], interaction of soil fauna and microflora determines soil dynamics. The contribution of their activity for formation of humus is an index for soil fertility. Bhatnagar [78] had expressed that at 20 to 40 cm depth in drillosphere zone there were 40% aerobic N-fixers, 13% anaerobic N-fixers, and 16% of denitrifiers. He attributed low C/N ratio in soils rich in earthworm population because of stimulation of N-fixers in drillosphere. Drillosphere provides necessary substrate for growth and establishment of microorganisms.

Recent developments in the country as well as at the global level are the application of detritivorous epigeic earthworms for organic manure/ vermicompost production from biodegradable organic materials recovered from agricultural lands, agro-based industries, and municipal solid waste. This field of study is closely associated with earthworm microbe interaction. The quality of the manure or vermicompost depends on micrcorganisms associated with the process of decomposition. Bhat [79] had reported that the diet formulation or the composition of organic matter used as feed influences the microflora associated with earthworm activity. Similar studies were made on enhanced N-fixers activity on using 2% neem cake in the feed mix of earthwormEudrilus eugeniae [80] (Table 9).

Table 9. Microbial population in neem cake enriched vermicompost [80].

Microbial population no./g vermicompost	Vermicompost with 2% neem cake	Vermicompost without neem cake
Fungi no. × 10^4	22.3	5.2
Bacteria no. × 10^6	15.0	7.8
Nitrogen fixers no. × 10^5	54.1	6.6

During winter months in Himalayan region, fungal population was higher in vermicomposting system than in the native soil [81]. Maintenance of temperature in vermicomposting system at a favourable level for earthworm activity might have been the reason for establishment of fungal population. Press mud, a by-product of sugar industry, is often used as one of the substrates in vermicomposting. Subjecting of this material to earthworm activity along with other organic matter has resulted in changes in microbial populations [82]. Rajani et al. [83] have related the microbial density and enzyme activity as a measure to assess the effectiveness of process of vermicomposting. It is essential to make an in-depth study to understand the mutualistic association between microflora and earthworms in mechanism of decomposition of organic matter. An increase in actinomycetes population was observed in the gut region of earthworms. Some of the isolates from gut region of earthworms have expressed growth stimulatory effect when used in pot cultures of tomato and finger millet [84].

The colony forming units (CFUs) of bacteria and fungi in the casts of P. corethrurus significantly deviated from the CFU found in adjacent soil. The correlation between the physicochemical parameters and microbial populations of the casts of P. corethrurus showed that the establishment of microbial population requires optimum moisture, organic carbon, and nitrogen content [20]. The vermicasts of P. ceylanensis showed 14 different fungal species belonging to the genera, Aspergillus, Chaetomium, Cladosporium, Cunninghamella, Fusarium, Mucor, Penicillium, and Rhizopus. Total nitrogen, phosphorus, potassium, calcium, copper, iron, and zinc were higher in vermicasts than in control (substrate without earthworms) while organic carbon and C/N ratio were lower in vermicasts. The total organic carbon was 42.3% in the control whereas it was 35.2% in the vermicasts of P. ceylanensis. The incubation of vermicasts (45 days) showed significant correlation with that of the increase in fungal population ($r = 0.720$; $P < .05$) and decrease in moisture content ($r = -0.984$; $P < .001$), and the decrease in moisture content statistically had no effect on the total fungal population in the vermicasts of P. ceylanensis [85]. The total microbial population, namely, bacteria, fungi, and actinomycetes was found to be many-fold higher than in the initial vermibed substrate and in substrate without earthworms (control). The initial count of bacteria, fungi, and actinomycetes in the control was 123.42 CFU $\times 10^7$ g^{-1}, 159.64 CFU $\times 10^3$ g^{-1}, and 86.90 CFU $\times 10^4$ g^{-1} whereas in castings (vermicompost) of P. ceylanensis the reported microbial populations were 268.62, 223.39, and 141.09 [86]. These observations clearly indicate the importance of microorganisms associated with earthworms in creating suitable environment for the standing crops as well as for vermicomposting of different organic wastes. It is still at the infancy to draw any inference regarding earthworm, microbe, and plant association.

Studies are also in progress to assess the inhibitory effects of the principles present in the body wall, gut extract, and of coelomic fluid on some selected plant and animal pathogens. The studies are at preliminary stages and it will require some more time to draw any conclusions based on the available data. Such interdisciplinary applications of earthworm research help to understand the functional complexity of these organisms other than their contribution to management of organic biodegradable residues as the major secondary detritivorous group.

8. EARTHWORMS AS BIOINDICATORS

Earthworms can also serve as indicators of several changes/factors associated with soil. Many studies clearly showed that the earthworms are best indicators of heavy metals, toxic pollutants, and direct and indirect anthropogenic changes in soil [87–89]. A study conducted in northern semiarid region of India showed the presence of earthworms to the maximum level wherever the farmers followed integrated farming (100%) practice and this was followed by organically managed (70%) and conventional (18.9%) agro-ecosystems. The earthworm abundance was directly related to the management practices and the values of ecological indices like Shannon diversity (H⍰), species dominance (C), the species richness (S), and evenness (E). This clearly illustrates the anthropogenic pressure on earthworm communities in arable lands [90]. Similar report from Ivory Coast is available on the impact of land-use changes and land-use intensification on earthworm populations and diversity in intermediate-disturbed systems [91]. Even though these studies suggest the use of earthworms as bioindicators of man-made changes, it necessitates more field and laboratory investigations to find out earthworm community structure, species interrelations, and the most efficient species to be used in biomonitoring of ecosystem degradation due to anthropogenic activities in the forest areas.

Certain toxic substances in soil affect the behaviour and physiology of earthworms that can serve as biomonitoring tool for their systematic effect on soil organisms and other higher organisms. For example, the presence of tetra ethyl lead (TEL) in leaded gasoline and lead oxide has a significant effect on behaviour, morphology, and histopathology of earthworms. Absorption of TEL into the tissues of earthworms produced severe effects, rupture of the cuticle, extrusion of coelomic fluid, and inflexible metameric segmentation. This led to desensitization of the posterior region and its fragmentation [92].

The efficient potential of earthworms in bioaccumulation of heavy metals in their tissues serves as ecological indicator of soil contaminants. As per the recent report from India, the level of DTPA extractable metals in casts of earthworms, Metaphire posthuma (endogeic) and Lampito mauritii (anecic) collected from cultivated land, urban garden and sewage soils were higher than those of surrounding soil. The concentration of Zn, Fe, Pb, and Mn in earthworm casts was higher in sewage soil followed by cultivated land and urban garden, respectively. There exists a close relationship between metal concentration in earthworm tissues and surrounding soils. The study also revealed the presence of species-specificity in metal accumulation in earthworms. Higher level of metal concentrating in the tissues was found in endogeic M. posthuma than in tissues of anecic L. mauritii. The difference in burrowing patterns may influence the patterns of bioaccumulations of metals apart from other contributory factors. Further, more detailed study is still required to elaborate the proposed hypothesis [93]. Analogous study conducted in Egypt also suggests that the variation in heavy metal concentration in soil and earthworms in different sites may be significant depending on soil properties and pollution status [88]. Sizmur and Hodson [94] evidently suggested that earthworms increase metal mobility and availability but more studies are required to determine the precise mechanism for this. So, this field of research with earthworm requires in depth research to understand the functional role of earthworms as bioindicators and bioconcentrators.

9. EARTHWORMS AND VERMICOMPOSTING: INDIAN SCENARIO

The familiar earthworm species, Eudrilus eugeniae, Eisenia fetida, Lumbricus rubellus, and Perionyx excavatus, are well known for their efficiency in vermicomposting. It is desirable to know about other species of earthworms that may be as efficient or better in their performance over the above mentioned species in a country having rich diversity of fauna for in situ and ex situ vermiculture. There are more than a dozen of earthworm species that have been reported to be efficient in vermicomposting. Most of the species that are included under genus Perionyx show great potential to work on organic matter. Apart from the well-known P. excavatus, other Perionyx species such as P. ceylanensis, P. bainii, P. nainianus, and P. sansibaricus are recently considered to be the potential vermicomposting earthworms [20, 95–97]. Future investigations provide scope for identifying more species with vermicomposting potential.

In natural systems, if earthworms are ecosystem engineers, in man made seminatural systems of organic residues, the detritivorous earthworms are saviors of biosphere from organic pollutants. From the review, it is very clear that the earthworm ecology needs much attention with reference to their functional role in different ecosystems. By the way of exploration, it might be possible to understand the significant role of earthworms in plant-microbe interactions. With regard to vermiculture, it is necessary to work on the idea of developing the consortia of earthworm species for vermiculture practices in India. It is always better to develop and encourage polycultures rather than maintaining monoculture. Moreover, with diversity in agricultural residues and by-products from agroindustries, it is essential to identify earthworms that will accept these materials with minimum effort and investment.

There are more than 500 species of earthworms distributed in different geographical regions in India, in different ecosystems. Being partly subtropical and partly temperate, majority of earthworms are endogeic or geophagous. Even among the epigeic earthworms (ca. 8%), those that are voracious feeders, are efficient biomass producers, and have short life cycle, high rate of fecundity, and high rate of adaptability to changing physicochemical properties of feed material can only serve as successful species for vermiculture. One has to look for these characters before recommending any species for vermiculture. The species that is promising under protected laboratory conditions in a small scale may fail to perform under field conditions when it is expected to work on large amount of organic matter. The present scenario in India shows that there is good response from the farmers to adopt the technology for producing vermicompost to use as soil amendment. They are reaping the benefits of using the recommended species for producing required quantity of vermicompost to fulfill the needs of their land and also to market the production to other neighbourhood farmers. Still many avenues remain open for the scientists to carry out research in this field to unravel various problems associated with the technology.

ACKNOWLEDGMENT

The author, Dr. N. Karmegam, sincerely acknowledges the authorities of Gandhigram Rural University, Gandhigram for giving permission to publish part of his Ph.D. thesis.

REFERENCES

1. K. E. Lee, Earthworms: Their Ecology and Relationships with Soils and Land Use, Academic Press, Sydney, Australia, 1985.

2. P. Lavelle, "The structure of earthworm communities," in Earthworm Ecology, J. E. Satchell, Ed., pp. 449–466, Chapman and Hall, London, UK, 1983.

3. J. E. Satchell and D. G. Lowe, "Selection of leaf litter in Lumbricus terrestris," in Progress in Soil Biology, O. Graff and J. E. Satchell, Eds., pp. 102–119, North Holland, Amsterdam, The Netherlands, 1967.

4. S. R. Ganihar, "Nutrient mineralization and leaf litter preference by the earthworm Pontoscolex corethrurus on iron ore mine wastes," Restoration Ecology, vol. 11, no. 4, pp. 475–482, 2003.

5. R. D. Kale and R. V. Krishnamoorthy, "Litter preferences in the earthworm Lampito mauritii,"Proceedings of Indian Academy Sciences, vol. 90, pp. 123–128, 1981.

6. A. V. Tiunov and S. Scheu, "Carbon availability controls the growth of detritivores (Lumbricidae) and their effect on nitrogen mineralization," Oecologia, vol. 138, no. 1, pp. 83–90, 2004.

7. M. C. Dash and V. C. Patra, "Density, biomass and energy budget of a tropical earthworm population from a grassland site in Orissa, India," Revue d Ecologie Et de Biologie du Sol, vol. 14, pp. 461–471, 1977.

8. R. D. Kale and R. V. Krishnamoorthy, "Cyclic fluctuations and distribution of three species of tropical earthworms in a farmyard garden in Bangalore," Rev d' Ecolgie et de Biol du Sol, vol. 19, pp. 67–71, 1982.

9. M. V. Reddy, V. P. K. Kumar, V. R. Reddy, et al., "Earthworm biomass response to soil management in semi-arid tropical Alfisol agroecosystems," Biology and Fertility of Soils, vol. 19, no. 4, pp. 317–321, 1995.

10. G. Tripathi and P. Bhardwaj, "Seasonal changes in population of some selected earthworm species and soil nutrients in cultivated agroecosystem," Journal of Environmental Biology, vol. 25, no. 2, pp. 221–226, 2004.

11. N. Karmegam and T. Daniel, "Effect of physico-chemical parameters on earthworm abundance: a quantitative approach," Journal of Applied Sciences Research, vol. 3, pp. 1369–1376, 2007.

12. A. Sathianarayanan and A. B. Khan, "Diversity, distribution and abundance of earthworms in Pondicherry region," Tropical Ecology, vol. 47, no. 1, pp. 139–144, 2006.

13. R. D. Kale, K. Bano, and R. V. Krishnamoorthy, "Anachoresis of earthworms," Bombay Natural History Society, vol. 78, pp. 400–402, 1981.

14. S. Subler, C. M. Baranski, and C. A. Edwards, "Earthworm additions increased short-term nitrogen availability and leaching in two grain-crop agroecosystems," Soil Biology and Biochemistry, vol. 29, no. 3-4, pp. 413–421, 1997.

15. C. H. Robinson, P. Ineson, T. G. Piearce, and A. P. Rowland, "Nitrogen mobilization by earthworms in limed peat soils under Picea sitchensis," Journal of Applied Ecology, vol. 29, no. 1, pp. 226–237, 1992.

16. C. A. Edwards and J. R. Lofty, "Nitrogenous fertilizers and earthworm populations in agricultural soils,"Soil Biology and Biochemistry, vol. 14, no. 5, pp. 515–521, 1982.

17. D. P. Knight, P. W. Elliott, J. M. Anderson, and D. Scholefield, "The role of earthworms in managed, permanent pastures in Devon, England," Soil Biology and Biochemistry, vol. 24, no. 12, pp. 1511–1517, 1992.

18. M. J. Shipitalo and K. R. Butt, "Occupancy and geometrical properties of Lumbricus terrestris L. burrows affecting infiltration," Pedobiologia, vol. 43, no. 6, pp. 782–794, 1999.

19. M. V. Reddy, V. R. Reddy, P. Balashouri, et al., "Responses of earthworm abundance and production of surface casts and their physico-chemical properties to soil management in relation to those of an undisturbed area on a semi-arid tropical alfisol," Soil Biology and Biochemistry, vol. 29, no. 3-4, pp. 617–620, 1997.

20. N. Karmegam and T. Daniel, "Selected physico-chemical characteristics and microbial populations of the casts of the earthworm, Pontoscolex corethrurus (Muller) and surrounding soil in an undisturbed forest floor in Sirumalai Hills, South India," Asian Journal of Microbiology, Biotechnology and Environmental Sciences, vol. 2, pp. 231–234, 2000.

21. J. M. Julka, R. Paliwal, and P. Kathireswari, "Biodiversity of Indian earthworms-an overview," inProceedings of Indo-US Workshop on Vermitechnology in Human Welfare, C. A. Edwards, R. Jayaraaj, and I. A. Jayraaj, Eds., pp. 36–56, Rohini Achagam, Coimbatore, India, 2009.

22. N. Karmegam, Studies on earthworms, vermiculture, vermicomposting and utilization of vermicompost for plant growth, Ph.D. thesis, Gandhigram Rural University, Gandhigram, Tamil Nadu, India, 2002.

23. N. Karmegam and T. Daniel, "A first report on the occurrence of a Megascolecid earthworm, Lampito kumiliensis (Annelida: Oligochaeta) in Sirumalai Hills of Tamil Nadu, South India," Ecology, Environment and Conservation, vol. 7, pp. 115–116, 2001.

24. B. G. M. Jamieson, "Preliminary descriptions of Indian earthworms (Megascolecidae: Oligocaheta) from the Palni hills," Bullettin du Muséum National d'Histoire Naturelle Section A, vol. 313, pp. 478–502, 1977.

25. S. A. Ismail and V. A. Murthy, "Distribution of earthworms in Madras," Proceedings of Indian Academy of Sciences (Animal Science), vol. 94, pp. 557–566, 1985.

26. N. Karmegam and T. Daniel, "Abundance and population density of three species of earthworms (Annelida: Oligochaeta) in foothills of Sirumalai (Eastern Ghats), South India," Indian Journal of Environment and Ecoplanning, vol. 3, pp. 461–466, 2000.

27. N. Karmegam and T. Daniel, "Population dynamics of a peregrine earthworm, Pontoscolex corethrurus in an undisturbed soil in Sirumalai

Hills of Tamil Nadu, South India," Journal of Ecological Research and Bioconservation, vol. 1, pp. 9–14, 2000.

28. T. Bhadauria and P. S. Ramakrishnan, "Population dynamics of earthworms and their activity in forest ecosystems of north-east India," Journal of Tropical Ecology, vol. 7, no. 3, pp. 305–318, 1991.

29. P. S. Chaudhuri and G. Bhattacharjee, "Earthworm resources of Tripura," Proceedings of National Academy of Sciences, India, vol. 69, pp. 159–170, 1999.

30. C. A. Edwards and P. J. Bohlen, Biology and Ecology of Earthworms, Chapman & Hall, London, UK, 1996.

31. S. A. Ismail, "Earthworm resources of Madras," in Proceedings of National Seminar on Organic Waste Utilization and Vermicomposting, Part B: Verms and Vermicomposting, M. C. Dash, B. K. Senapati, and P. C. Mishra, Eds., pp. 8–15, 1986.

32. K. Bano and R. D. Kale, "Earthworm fauna of Southern Karnataka, India," in Advances in Management and Conservation of Soil Fauna, G. K. Veeresh, D. Rajagopal, and C. A. Virakthamath, Eds., pp. 627–634, Oxford & IBH, New Delhi, India, 1991.

33. R. V. Krishnamoorthy, "Competition and coexistence in a tropical earthworm community in a farm garden near Bangalore," Journal of Soil Biology and Ecology, vol. 5, pp. 33–47, 1985.

34. T. Bhadauria and P. S. Ramakrishnan, "Earthworm population dynamics and contribution to nutrient cycling during cropping and fallow phases of shifting agriculture (jhum) in north-east India," Journal of Applied Ecology, vol. 26, no. 2, pp. 505–520, 1989.

35. S. A. Ismail, C. Ramakrishnan, and M. M. Anzar, "Density and diversity in relation to the distribution of earthworms in Madras," Proceedings of Indian Academy of Sciences, Animal Sciences, vol. 99, no. 1, pp. 73–78, 1990.

36. B. R. Kaushal, S. P. S. Bisht, and S. Kalia, "Population dynamics of the earthworm Amynthas alexandri(Megascolecidae: Annelida) in cultivated soils of the Kumaun Himalayas," Applied Soil Ecology, vol. 2, no. 2, pp. 125–130, 1995.

37. P. S. Chaudhuri, S. Nath, and R. Paliwal, "Earthworm population of rubber plantations (Hevea brasiliensis) in Tripura, India," Tropical Ecology, vol. 49, no. 2, pp. 225–234, 2008.

38. A. V. Evans and W. J. Mc. L. Guild, "Studies on the relationships between earthworms and soil fertility. IV. On the life cycles of some British Lumbricidae," Annals of Applied Biology, vol. 35, pp. 471–484, 1948.

39. B. K. Senapati and S. K. Sahu, "Population, biomass and secondary production in earthworms," inEarthworm Resources and Vermiculture, pp. 57–78, Zoological Survey of India, Calcutta, India, 1993.

40. S. R. Ganihar, "Earthworm distribution with special reference to physicochemical parameters,"Proceedings of Indian National Science Academy B, vol. 62, pp. 11–18, 1996.

41. G. González, X. Zou, and S. Borges, "Earthworm abundance and species composition in abandoned tropical croplands: comparisons of tree plantations and secondary forests," Pedobiologia, vol. 40, no. 5, pp. 385–391, 1996.

42. K. Auerswald, S. Weigand, M. Kainz, and C. Philipp, "Influence of soil properties on the population and activity of geophagous earthworms after five years of bare fallow," Biology and Fertility of Soils, vol. 23, no. 4, pp. 382–387, 1996.

43. P. Lavelle, "Earthworm activities and the soil system," Biology and Fertility of Soils, vol. 6, no. 3, pp. 237–251, 1988.

44. S. Nordström and S. Rundgren, "Environmental factors and Lumbricid associations in Southern Sweeden," Pedobiologia, vol. 13, pp. 301–326, 1974.

45. B. K. Senapati and M. C. Dash, "Influence of soil temperature and moisture on the reproductive activity of tropical earthworms of Orissa," Journal of Soil Biology and Ecology, vol. 4, pp. 13–21, 1984.

46. A. Kretzschmar and C. Bruchou, "Weight response to the soil water potential of the earthwormAporrectodea longa," Biology and Fertility of Soils, vol. 12, no. 3, pp. 209–212, 1991.

47. R. D. Kale and R. V. Krishnamoorthy, "Distribution and abundance of earthworms in Bangalore,"Proceedings of Indian Academy of Sciences, vol. 88, pp. 23–25, 1978.

48. C. Fragoso and P. Lavelle, "The earthworm community of a Mexican tropical rain forest (Chajul, Chiapas)," in On Earthworms, A. M. Bonvincini Paglai and P. Omodeo, Eds., Selected Symposia and Monographs U.Z.I., pp. 281–295, Modena, Mucchi, Italy, 1987.

49. N. Joshi and S. Aga, "Diversity and distribution of earthworms in a subtropical forest ecosystem in Uttarakhand, India," The Natural History Journal of Chulalongkorn University, vol. 9, pp. 21–25, 2009.

50. V. B. Tembe and P. J. Dubash, "The earthworms: a review," Journal of Bombay Natural History Society, vol. 58, pp. 171–201, 1961.

51. S. K. Roy, "Studies on the activities of earthworms," Proceedings of Zoological Society, Calcutta, vol. 10, pp. 81–98, 1957.

52. G. E. Gates, "Ecology of some earthworms with special reference to seasonal activity," American Midland Naturalist, vol. 66, pp. 61–86, 1961.

53. M. C. Dash and V. C. Patra, "Worm cast production and nitrogen contribution to soil by a tropical earthworm population from a grassland site from Orissa, India," Revue d Ecologie Et de Biologie du Sol, vol. 16, pp. 79–83, 1979.

54. L. Norgrove and S. Hauser, "Effect of earthworm surface casts upon maize growth," Pedobiologia, vol. 43, no. 6, pp. 720–723, 1999.

55. M. V. Reddy, "Annual cast production by the Megascolecid earthworm, Pheretima alexandri (Beddard),"Comparative Physiology and Ecology, vol. 8, pp. 84–86, 1983.

56. P. Lavelle, "Les vers de terre de la savane de Lamto. Analyse d'un Ecosysteme tropical humide: LA Savanede Lamto (Côte d'Ivoire)," Bullettin De Liaison des chercheurs de Lamto, vol. 5, pp. 133–136, 1974.

57. S. D. Nijhawan and J. S. Kanwar, "Physicochemical properties of earthworm castings and their effect on the productivity of the soil," Indian Journal of Agricultural Sciences, vol. 22, pp. 357–373, 1952.

58. A. K. Dutt, "Earthworms and soil aggregation," Journal of American Society for Agronomy, vol. 48, p. 407, 1948.

59. R. J. Swaby, "The influence of earthworm on soil aggregation," Journal of Soil Science, vol. 1, pp. 195–197, 1950.

60. J. N. Parle, "A microbiological study of earthworm casts," Journal of General Microbiology, vol. 31, pp. 13–22, 1963.

61. A. M. Habibullah and S. A. Ismail, "Preference to soil fractions and the effect of soil compaction on the casting and burrowing behaviour of the earthworm Lampito mauritii," Journal of Soil Biology and Ecology, vol. 5, pp. 26–32, 1985.

62. G. W. Agarwal, K. S. K. Rao, and L. S. Negi, "Influence of certain species of earthworms on the structure of some hill soils," Current Science, vol. 27, p. 213, 1958.

63. M. Puttarudraiah and K. S. S. Sastry, "A preliminary study of earthworm damage to crop growth," Mysore Agriculture Journal, vol. 36, pp. 2–11, 1961.

64. P. M. Fraser, M. H. Beare, R. C. Butler, T. Harrison-Kirk, and J. E. Piercy, "Interactions between earthworms (Aporrectodea caliginosa), plants and crop residues for restoring properties of a degraded arable soil," Pedobiologia, vol. 47, no. 5-6, pp. 870–876, 2003.

65. M. A. Callaham Jr. and P. F. Hendrix, "Impact of earthworms (Diplocardia: Megascolecidae) on cycling and uptake of nitrogen in coastal plain forest soils from northwest Florida, USA," Applied Soil Ecology, vol. 9, no. 1–3, pp. 233–239, 1998.

66. S. A. Materechera, O. T. Mandiringana, and K. Nyamapfene, "Production and physico-chemical properties of surface casts from microchaetid earthworms in central Eastern Cape," South African Journal of Plant and Soil, vol. 15, no. 4, pp. 151–157, 1998.

67. J. G. Shrikande and A. N. Pathak, "Earthworms and insects in relation to soil fertility," Current Science, vol. 17, pp. 327–328, 1948.

68. N. V. Joshi and B. V. Kelkar, "Role of earthworms in soil fertility," Indian Journal of Agricultural Science, vol. 22, pp. 189–196, 1952.

69. A. N. Ganeshamurthy, K. M. Manjaiah, and A. S. Rao, "Mobilization of nutrients in tropical soils through worm casting: availability of micronutrients," Soil Biology and Biochemistry, vol. 30, no. 13, pp. 1839–1840, 1998.

70. R. D. Kale and R. V. Krishnamoorthy, "The calcium content of body tissues and castings of earthwormPontoscolex corethrurus (Annelida-Oligochaeta)," Pedobiologia, vol. 20, pp. 309–315, 1980.

71. C. A. Edwards and N. Q. Arancon, "Interactions among organic matter, earthworms and microorganisms in promoting plant growth," in Soil Organic Matter in Sustainable Agriculture, pp. 327–376, 2004.

72. J. V. Bhat, S. R. Khambata, G. B. Maya, C. A. Sastry, R. V. Iyer, and V. Iyer, "Effect of earthworms on the microflora of the soil," Indian Journal of Agricultural Science, vol. 30, no. 2, pp. 106–114, 1960.

73. S. R. Khambata and J. V. Bhat, "A contribution to the study of the intestinal microflora of Indian earthworms," Archives of Microbiology, vol. 28, no. 1, pp. 69–80, 1957.

74. M. C. Dash, P. C. Mishra, and N. Behara, "Fungal feeding by a tropical earthworm," Tropical Ecology, vol. 20, pp. 9–12, 1979.

75. T. C. Dlamini, R. J. Haynes, and R. van Antwerpen, "Exotic earthworm species dominant in soils on sugarcane estates in the Eshowe area of the north coast of Kwazulu-Natal," in Proceedings of Annual Congress of South African Sugar Technologists Association, vol. 75, pp. 217–221, 2001.

76. G. G. Brown, I. Barois, and P. Lavelle, "Regulation of soil organic matter dynamics and microbial activity in the drilosphere and the role of interactions with other edaphic functional domains," European Journal of Soil Biology, vol. 36, no. 3-4, pp. 177–198, 2000.

77. A. Kretzschmar, "Importance of the interaction of soil fauna and microflora for formation of humus and development of organic substance," Berichte uber Landwirtschaft son der heft, vol. 206, pp. 117–126, 1992.

78. T. Bhatnagar, "Lombriciens et humification: un aspect nouveau de l'incorporation microbienne d'azote induite par les vers de terre," in Biodegradation et Humification, pp. 169–182, Pierron, Satreguemines, France, 1975.

79. J. V. Bhat, "Suitability of experimentation diets for earthworm culture," Current Science, vol. 43, pp. 266–268, 1974.

80. R. D. Kale, K. Vinayaka, K. Bano, and D. J. Bagyaraj, "Suitability of neem cake as an additive in earthworm fed and its influence on the establishment of microflora," Journal of Soil Biology and Ecology, vol. 6, pp. 98–103, 1986.

81. R. Nagar, N. Joshi, S. Dwivedi, and V. K. Khaddar, "A physicochemical and micofloral profile of vermicompost in Tarai region of Himalaya in winter season," Research on Crops, vol. 5, pp. 51–54, 2004.

82. K. Parthasarathi, L. S. Ranganathan, and J. Zeyer, "Species specific predation of fungi by Lampito mauritii(Kinb) and Eudrilus eugeniae (Kinb) reared on press mud," in Session 3. Earthworms and Vermicomposting, D. J. Bagyaraj, Ed., VII National Symposium on Soil Biology and Ecology, Ab, p. 69, 2001.

83. B. S. Rajani, D. Radhakrishna, V. R. Ramakrishnaparama, R. D. Kale, and A. N. Balakrishna, "Influence of vermicomposting of urban solid wastes on microbial density and enzyme activities," in Session 4, Soil Biota as Bioindicators of Soil Health, D. J. Bagyaraj, Ed., VII National Symposium on Soil Biology and Ecology, Ab, p. 69, 2001.

84. B. Raghavendra Rao, Assessment of microbial and biochemical quality of urban compost and its impact on soil health, Ph.D. thesis, University of Agricultural Sciences, Bangalore, India, 2001.

85. M. Prakash, M. Jayakumar, and N. Karmegam, "Physico-chemical characteristics and fungal flora in the casts of the earthworm, Perionyx ceylanensis Mich. reared in Polyalthia longifolia leaf litter," Journal of Applied Sciences Research, vol. 4, pp. 53–57, 2008.

86. M. Jayakumar, V. Karthikeyan, and N. Karmegam, "Comparative studies on physico-chemical, microbiological and enzymatic activities of vermicasts of the earthworms, Eudrilus eugeniae, Lampito mauritii and Perionyx ceylanensis cultured in press mud," International Journal of Applied Agricultural Research, vol. 4, pp. 75–85, 2009.

87. J. J. Hinton and M. M. Veiga, "Earthworms as bioindicators of mercury pollution from mining and other industrial activities," Geochemistry: Exploration, Environment, Analysis, vol. 2, no. 3, pp. 269–274, 2002.

88. H. M. Mahmoud, "Earthworm (Lumbricus terrestris) as indicator of heavy metals in soils," Online Journal of Veterinary Research, vol. 11, pp. 23–37, 2008.

89. C. B. Iwai, P. Yupin, and B. N. Noller, "Earthworm: potential bioindicator for monitoring diffuse pollution by agrochemical residues in Thailand," KKU Research Journal, vol. 139, pp. 1081–1088, 2008.

90. S. Suthar, "Earthworm communities as bioindicator of arable land management practices: a case study in semiarid region of India," Ecological Indicators, vol. 9, no. 3, pp. 588–594, 2009.

91. J. E. Tondoh, L. M. Monin, S. Tiho, and C. Csuzdi, "Can earthworms be used as bio-indicators of land-use perturbations in semi-deciduous forest?" Biology and Fertility of Soils, vol. 43, no. 5, pp. 585–592, 2007.

92. J. Venkateswara Rao, P. Kavitha, and A. Padmanabha Rao, "Comparative toxicity of tetra ethyl lead and lead oxide to earthworms, Eisenia fetida (Savigny)," Environmental Research, vol. 92, no. 3, pp. 271–276, 2003.

93. S. Suthar, S. Singh, and S. Dhawan, "Earthworms as bioindicator of metals (Zn, Fe, Mn, Cu, Pb and Cd) in soils: is metal bioaccumulation affected by their ecological category?" Ecological Engineering, vol. 32, no. 2, pp. 99–107, 2008.

94. T. Sizmur and M. E. Hodson, "Do earthworms impact metal mobility and availability in soil?—a review,"Environmental Pollution, vol. 157, no. 7, pp. 1981–1989, 2009.

95. S. Suthar, "Growth and fecundity of earthworms: Perionyx excavatus and Perionyx sansibaricus in cattle waste solids," The Environmentalist, vol. 29, no. 1, pp. 78–84, 2009.

96. N. Karmegam and T. Daniel, "Investigating efficiency of Lampito mauritii (Kinberg) and Perionyx ceylanensis Michaelsen for vermicomposting of different types of organic substrates," The Environmentalist, vol. 29, no. 3, pp. 287–300, 2009.

97. N. Karmegam and T. Daniel, "Growth, reproductive biology and life cycle of the vermicomposting earthworm, Perionyx ceylanensis Mich. (Oligochaeta: Megascolecidae)," Bioresource Technology, vol. 100, no. 20, pp. 4790–4796, 2009.

Chapter 8

Are Australian and United States Farmers Using Soil Information for Soil Health Management?

Lisa Lobry de Bruyn [1,*] and Susan Andrews [2]

[1]School of Environmental and Rural Science, University of New England, Armidale 2351, Australia

[2]United States Department of Agriculture, Natural Resources Conservation Service, Denton, NE 68339, USA

ABSTRACT

Soil health is an essential requirement of a sustainable, functioning agroecosystem. Tracking soil health to determine sustainability at the local level largely falls to farmers, even though they often lack access to critical information. We examine farmers' participation in gathering soil information at the farm and paddock scale over the last two decades in Australia and the United States, by reviewing national-level reporting of farmer use of soil testing and farm planning as well as qualitative research on farmer perspectives. The level of participation in soil testing and farm planning has remained stable in the last two decades, with only 25% and 30% of landholders, respectively, participating nationally, in either country. The review revealed national-level reporting has a number of limitations in understanding farmers' use of soil information and, in particular, fails to indicate the frequency and intensity of soil testing as well as farmer motivation to test soil or what they did with the soil information. The main use of soil testing is often stated as "determining fertilizer requirements", yet data show soil testing is used less commonly than is customary practice. In Australia and in the United States, customary practice is three and half times more likely for decisions on fertilizer application levels. The rhetoric is heavy on the use of soil testing as a decision tool, and that it guides best practices. However, given that only a quarter of farmers are soil testing, and doing so infrequently and in low densities, the level of information on soil health is poor. While farmers report consistent monitoring of soil conditions, few have consistent records of such. In contrast to the information on the poor state of soil health, there is strong farmer interest in procuring soil health benefits through changes in farm practices such as conservation tillage or cover crops, even if they are unable to demonstrate these soil health benefits through soil testing. Many farmers report the use of observation in lieu of laboratory testing. Finally, we point to the need for soil information to include observational indicators to best allow a blend of traditional extension strategies with digital technology to create communities of interest in soil management. This would transcend the boundaries between those with expertise and those with experience in soil health management.

KEYWORDS

soil health; soil monitoring; soil information; soil testing; farmers

1. INTRODUCTION

Soil health or quality is critical to well-functioning ecosystems that can provide a number of services [1,2]. For more than two decades, scientists have called for the broader recognition of the multi-functionality of soils [3,4,5], a move not only beyond the narrow focus on productivity but also beyond erosion [6]. The importance of on- and off-site impacts of soil health and its relationship to sustainability was recognized in the US by the National Research Council, some 22 years ago [7]. The concept of soil quality (or health) is defined as "the capacity of a soil to function, both within its ecosystem boundaries and with the environment external to that ecosystem" [8]. Changes in soil health occur at the human time scale, in response to human-induced and natural disturbances [9]. Land degradation and loss of soil health is a worldwide issue of concern [10]. A recent assessment of the status of the world's soil stated that effective responses to deleterious soil change must be derived from policies that direct attention to the activities of individual land managers [11].

Soil health at national and regional scales is often measured more by the extent of land degradation than by specific and sometimes subtle changes in soil properties. This suggests a lack of specific soil data to determine the state of the soil resource and drivers of change. Without soil health monitoring, the state or trend line of the soil resource is unknown, and the likelihood of further deterioration is possible. There are a number of cyclical programs in Australia (AU) and in the United States (US) which document the status of and threats to soil functioning, such as erosion, salinization, soil acidity and loss of soil carbon at the regional and national level (e.g., the State of Environment reporting, conducted nationally every five years, and the National Land and Water Audit in 1998–2002 in AU, as well as the National Resources Inventory (NRI), conducted every five years for water and wind erosion in the US). However, the imperative to monitor soil, especially at the landholder level, has not gained momentum, despite the bleak outlook and continuing soil degradation [12,13]. A 2015, FAO report states, "*There is a qualitative appreciation of the pressures on soil resources but limited consistent evidence on their condition or trajectories of change*" [11], furthering the case for monitoring.

Despite the recognition of soil health as a critical variable in ecosystem change by many academics [14,15], there are a number of barriers to collecting and acting on soil information, including a lack of resources (monetary and human) at multiple scales, a lack of education to understand the need, and privacy issues associated with sharing information. This paradox is also at odds with a growing grassroots interest in soil health, in both AU and the US, which is only partially supported by government programs, and often sits outside traditional institutions.

Both the US and AU have a long history of dealing with soil quality degradation [16,17]. In AU, farm planning and soil testing are viewed as voluntary approaches to drive a change in attitude and increase adoption rates of best management practices (BMPs). Conservation is sought via capacity-building through training and altering ways of thinking [18]. The focus has been on extension, building capacity and raising awareness of the issues with landholders. Since the 1990s, funding was through the National

Landcare Program (1992–2008). From 2008 to the present, Caring for Our Country Program filled this role [16]. In the US, conservation is also voluntary. The role of extension varies state by state. In addition, the US Farm Bill contains a section on conservation. Much of the US activity is mandated by the Farm Bill, which legislates voluntary conservation programs, such as the Environmental Quality Improvement Program (EQIP, established in 1996) and the Conservation Security Program (2002–2008), renamed the Conservation Stewardship Program (2009-present). These programs offer cost-sharing for installation or implementation of natural resource conservation practices or an annual payment based on practices already in place plus additional practices the landowner agrees to begin during the five- to 10-year contract. The United States Department of Agriculture, Natural Resources Conservation Service (USDA NRCS) (established in 1985), offers conservation technical assistance and financial cost-sharing and incentive payments for these programs. In addition, extension services and grant programs are coordinated through the USDA National Institute of Food and Agriculture (NIFA), previously the Cooperative State Research, Education, and Extension Service (CSREES).

We review developments, over the last two decades, in Australian and US soil health information and soil building practices by examining farmers' soil data collection and their perspectives on soil health. Following the logic that soil testing is a "decision aid" that can provide the farm operator with information needed to pursue farming practices that moderate chemical use, we use soil testing and farm planning as a way to gauge farmers' use of soil information for soil health management. Our premise is that access to and use of good quality, local soil information, including identification of soil types and their soil health status (here using available soil testing as a proxy), will enable farmers to determine the best course of action to either improve or maintain soil health. We examine whether these tools have been used to guide soil health management. Specially, we determine, for AU and the US:

1) the amount of farmer participation that exists in soil testing and farm planning;

2) the nature of soil testing and farmer motivation to capture information on soil health;

3) whether farmers use soil testing for decision-making, including the adoption of soil building practices;

4) limitations to widespread use of soil information and soil testing for soil health management; and

5) approaches to improving farmers' access to soil information, and its potential use for the adaptive management of agroecosystems.

2. APPROACH

We examined national records from both AU and the US, reviewed the available literature on farmer use of soil information, and supported our investigations with interviews of scientists and practitioners, as well as focus group results. We also made one critical assumption: available soil test information is thorough enough to serve as a proxy for (at least some) indicators of soil health. We understand that soil testing has historically focused on chemical indicators of soil health, such as soil pH and macronutrients, while largely ignoring important physical and biological indicators [19]. Now that organic carbon is more widely available and affordable in most laboratories, we believe our assumption is more reasonable. Readily available soil testing in both countries does allow detection of many soil health issues, such as soil acidity or alkalinity, loss of major nutrients, toxicity and changes in organic carbon [20,21,22,23,24].

We data-mined a number of databases in Australia: the Australian Bureau of Agricultural Resource Economics and Sciences (ABARES) [25]; the Australian Bureau of Statistics (ABS) [26] (Figure A1); and documents produced by the Federal Government of Australia, including ABARES and ABS. We reviewed the data collected from: ABARES surveys, as part of the*Farms Surveys* series (1988–1989, 2001–2002, 2004, 2006 and 2009); ABS survey data from the Agricultural Resource Management Survey (ARMS); the Land and Management Practices Survey (LAMP); and the Rural Environment and Agricultural Commodities Survey (REAACS) (from 2007 to 2013). We examined soil testing, farm planning and conservation practice application. In the United States (US), the commodity chosen for assessing the use of soil information by farmers was corn due to its dominance as a crop in terms of land area and fertilizer use (Figure A2). We used publications produced from the United States Department of Agriculture, the Economic Research Service (USDA, ERS), the Agricultural Resource Management Study (ARMS) Corn Crop Production Practices and Cost Report Phase 11 Survey and ARMS data online from 1996 to 2010 [27]. A number of USDA National Agricultural Statistical Service (NASS) ARMS data from 2000, 2005 and 2010 were accessed, which examined the areas specifically planted with corn that were subject to soil testing and fertilizer application. Tailored reports were provided by Jodie Mewett (ABARES) and Rachael Brown (USDA, ERS) for data not available online or where the sampling population could not be determined online. All Figures and Tables have been constructed by authors from the data sources as cited. We also contacted a number of the authors of these reports to delve further into their methodology and data sets, as well as garner their opinions on the current situation on soil information gathering and use by farmers.

In addition, throughout this review, results of farmer focus groups with US farmers in California (CA) and Illinois and Iowa (IL-IA) (held at the border of the two states) are reported. Some of these CA focus group results, particular on farmer preferences for the presentation of soil quality information, along with a full methodology, were published by Andrews *et al.*[28]. The IA-IL focus groups were funded by USDA Sustainable Agriculture Research and Education (SARE) and culminated in the design and development of an educational website [29]. At each location, in the northern and southern Central Valley of California and in the Quad City area at the border of Iowa and Illinois, four focus groups of farmers and other agricultural professionals, with approximately 10 participants, were held in 2000–2001. Only the farmer focus groups' results are reported here (three groups in total). Focus groups were asked a series of pre-planned, open-ended questions. The questions varied somewhat between CA and IL-IA due to different overall objectives for the studies. However, many questions were similar enough to allow for combined analysis. Efforts were made to hear from all participants. Responses were recorded and transcribed. Themes were identified using NVivo® v.1.2 software (QSR International, Melbourne, Australia) qualitative analysis software (Note: reference to trade names and companies is made for information purposes only and does not imply endorsement by the USDA or the University of New England). Response themes and supporting quotes are presented throughout the review and discussed in relation to other published survey information.

Although this qualitative data was gathered approximately 15 years ago, this part of the information has not been published previously. Due to the lack of major changes in data use and policy during this time and because of the similarity to more recent farmer comments garnered in less rigorous ways, we assume that the information is still relevant today. Additionally, this study falls within the two-decade time frame for this review.

3. FARMER PARTICIPATION IN SOIL TESTING AND FARM PLANNING

3.1. Soil Testing Trends

Using the REAACS data, we found a trend toward reduced soil testing, particularly among grazing enterprises, in AU from 2007 to 2013 (Figure 1). The figure represents the percent of farm businesses performing soil tests for pH and nutrients (Figure 1) by extracting them from an online data portal, called Monitor, and by requesting specific adjustments, made by Jodie Mewett, to include population size and 2012–2013 data [25].

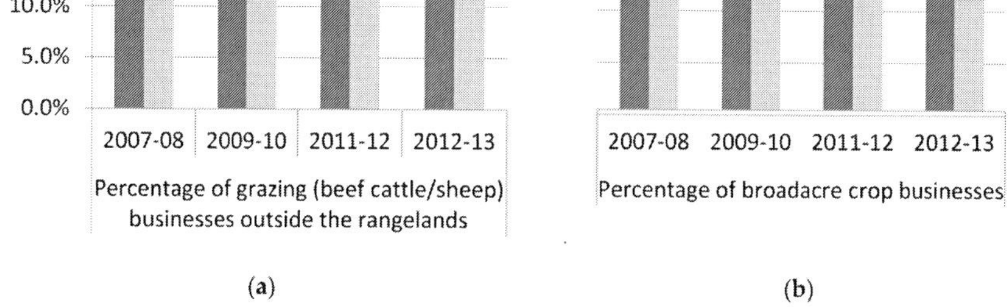

Figure 1. In Australia, (**a**) the percentage of grazing and (**b**) broadacre businesses performing soil tests for soil pH and nutrients in previous financial year (Data source: [30,31,32,33]).

Data collected by ABS (2007–2013) on soil testing indicates only whether the practice occurred in the previous year (Figure 1). The question response options were either, "Yes", "No" or "Not applicable", except when the ABS data were publicly released, and then the responses to the soil testing question were reported as the proportion who indicated "Yes" they were soil or plant testing, and none of the other responses. Also, by the question specifying the period of soil testing as the previous year it could be underestimating the participation of farmers whose soil testing was conducted outside that period. To illustrate the issue, we found that one researcher asked farmers whether they did soil testing in the last five years. About 65% of respondents indicated they had, while 28.5% had not [34]. The responses led the researcher to write "...suffice to say that soil testing is not a common undertaking ..." [35]. Nevertheless, this response rate is more than double that reported by ABS. We strongly suggest that these survey questions be rephrased to capture soil testing occurring at longer-than-annual intervals. In surveys of grain growers, landholders were asked about soil testing frequency. Results showed approximately 30% of grain farmers annually test soils [36, 37]. An additional 37% of grain farmers stated they routinely did soil testing every two or three years [36,37]. This means that the remaining 33% of grain farmers were either soil testing at an interval greater than three years or not at all.

ABS surveys also do not provide information about the motivation, nature or intensity of soil testing. In earlier versions of this survey (pre-2007), landholders were asked to respond to the statement: "*Which of the following farm practices are part of your farm management program, and for the wheat-sheep and high rainfall zone farms only—soil or plant tissue test to determine fertilizer or soil conditioner requirements.*" However, this statement raises other issues in that it was not time-bound, and did not specifically ask when soil testing occurred, only if it was part of their farm management practice. Additionally, the motivation behind the soil testing is assumed in the wording of the statement. It is possible that any farmers performing soil tests for other reasons, such as soil monitoring, could answer in the negative. In post-2007 surveys the ABS question on soil testing was altered to "*Have you undertaken any soil testing on this holding between [financial years]?*" with a Yes/No response, followed by "*What did the soil testing include? (Please tick all that apply): pH, Nutrient levels, Organic Matter levels, Salinity (E.C.), Other.*" So in the post-2007 survey instrument there was some indication of the type of soil tests being performed, but again no idea of sampling intensity or the farmer's motivation to undertake testing. Finally, in the most recent ABS survey on Land Management and Farming 2013–2014, the question on soil testing is no longer asked [38].

From the 2001 US ARMS survey, Lambert *et al.* [39] determined that less than 30% of US corn farmers were soil/tissue testing, and even fewer (under 20%) were using available soil maps. In fact, the testing of soil nitrogen (N), phosphorus (P), and potassium (K) has declined since 1996, despite the relatively constant fertilizer application area and amounts of fertilizer applied (Figure 2). The examination of three reports for 1996, 2001 and 2010 [40,41,42] on characteristics of corn production costs showed that N fertilizer was applied to almost 100% of the planted area. Amounts have remained steady at an average of 158 kg N/treated hectare with corn accounting for 46% of the US fertilizer consumption in 2010 [43]. In the 1996 ARMS survey of the US Corn Belt states, an area comprising over half of all corn producers and producing 70% of all US corn, only 14% of farms tested soil for N, with just over half of those farms using the recommended level [41]. ARMS surveys (conducted every five years), like the AU surveys, by confining responses to soil testing to the previous or current year in a particular field, could be underestimating participation in soil testing for those farmers not soil testing during that period.

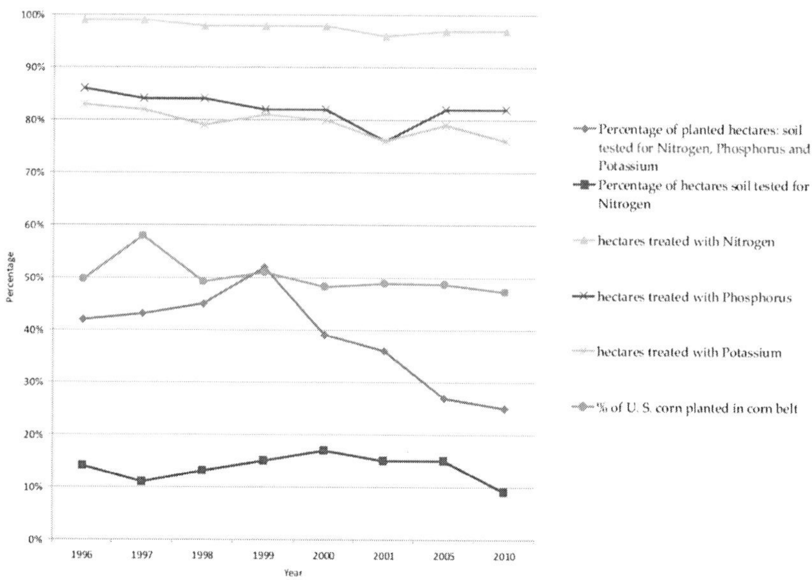

Figure 2. The percentage of US Corn belt hectares tested for nitrogen, phosphorus and potassium estimated from ARMS surveys in 1996 to 2010. (Data source: [27]).

In Australia, despite a five-fold growth in the number of soil samples analyzed from the early 1990s, it is still a tool used at low intensities. The estimates in Figure 3 were based on the most likely farmer participation rates of 30% of the known cropping area for that year, calculated from ABS statistics, and equate to approximately five soil samples per 100 ha. In 1998, Plank [44] also showed that the volume of soil samples analyzed in the US for commercial purposes was 2 M in 1996, but given that the US has 2.22 M farms [45], the number of soil samples per farm is probably quite low. The IPNI report for 2000–2001 also indicated highly variable sampling density across the US with, on average, one sample per 67 ha, ranging from one sample per 9 ha in Georgia to one sample per 1214 ha in Wyoming [46]. Placing recent US figures on soil testing volume [47,48] through commercial laboratories, and assuming they are only applied to cropland, the number of soil samples per 100 ha is greater than in Australia. However, at 10 soil samples per 100 ha (with 30% farmer participation), it is still not at a high density [45]. The trends in soil testing show the number of farmers involved is not growing even though the volume of soil tests analyzed seems to indicate it has grown (Figure 1, Figure 2 and Figure 3).

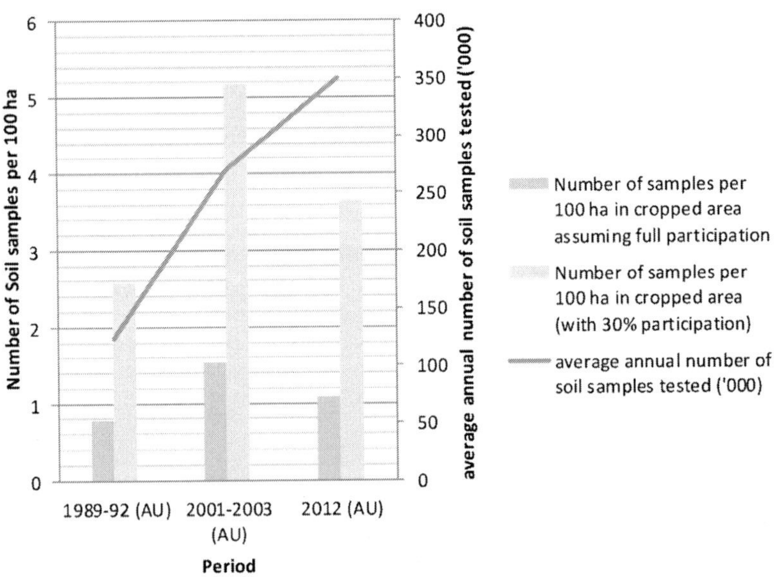

Figure 3. Changes in soil sampling number and intensity from 1989 to 2012 in Australia, from private soil testing laboratories. Data sources: [21,23, 33,49,50,51,52].

Another way of examining this data is to hypothesize why farmers appear to be doing only limited soil testing, both in terms of frequency and intensity of sampling. One soil scientist interviewed suggested that by avoiding soil testing, farmers were also avoiding the need to address the declining soil condition [53]. This assumes that some farmers either do not foresee the long-term benefits of building soil health or have economic pressures that do not allow them to think beyond the short term. Farmers in the US focus groups, seemingly in agreement with the latter reason for this avoidance behavior, identified economic pressures as a deterrent to performing soil building practices.

"I think the more and more farming becomes tougher economically, then everybody continues to look shorter and shorter term. ... So, sometimes that limits what you think might be the optimum, long-run approach."

—CA farmer

Soil testing could become expensive and time-consuming for farmers, especially if undertaken at the frequency and intensity required for monitoring soil health.

3.2. Farm Planning and Conservation Plans

Tabulating the number of written farm plans in AU reveals that the percentage of landholders involved has remained static in recent years at 20% [54] but has declined from the levels recorded in the early 1990s (36% and 30%, respectively) [55,56] (Table A1). Hodges and Goesch [57], in their 2004–2005 ABARE survey of Australian farms, found that just over 20% of farmers have a farm plan. Those plans included details on production activities (80%), natural resource management (75%), business activities (63%), people management (30%) and references to regional NRM priorities (25%). This data showed that soil and plant testing was part of the farm management program for 59% of the surveyed farmers across all Australian farms in 2004–2005 [57]. Involvement in a natural resource management initiative (e.g., National Landcare Program, National Heritage Trust, National Action Plan for Salinity and Water Quality) seems to correlate with a higher adoption or use of sustainable land management practices, which includes the monitoring of pasture conditions (15% higher than NRM non-participants), soil and plant tissue tests (24% higher than non-participants) and monitoring of water tables (10% higher) [57]. For those landholders involved in Landcare activities in 2004–2005, an overwhelming majority (89%) stated they learned about the causes, signs and treatment of land degradation. Similarly, 82% said they learned skills to monitor degradation or implement conservation practices on ground works [54]. However, the data is unclear about what types of degradation and monitoring schemes they were informed by. The influence of Landcare on farmer participation in soil testing is not likely to be as significant given that recent Landcare membership is under 20% Australia-wide compared with a peak in membership in at 41%, (2001–2002) [25]. The programs available to support farmers to write a farm plan are also diminished in their activity, as publicly funded schemes have become outsourced to private enterprise or fee-for-service providers such as education providers, including Profarm courses run by Tocal College in New South Wales, AU [58].

In contrast to the 2004–2005 ABARE report [57], their 2012 report on drivers of practice change tallied the percentage of landholders with a written property plan at 31% Australia-wide, an 11% increase in just seven years [59] (Table A1). The later report also found that property management plans were more focused on the financial side of the business (76%) or farm production activities (77%) than on the management of natural resources (51%), which seems to indicate a 24% decline when compared with the previous report [59]. However, the margins for error of these reports were not available. When farmers were asked how frequently they refer to their planning documents when making a decision, 35% of farmers responded that they "always" refer to their plans when it comes to farm financial decisions, while only 14% of farmer respondents "always" refer to their plans for natural resource decisions [59]. Therefore, if consultation of such plans is minor, then the guidance they provide over soil health management is also small.

To receive financial assistance for conservation on US working lands, land managers must meet a set of eligibility requirements. To qualify for conservation programs promulgated by USDA NRCS, including Conservation Stewardship Program (CSP) and Environmental Quality Improvement Program (EQIP), farmers are expected to have a conservation plan, which can be completed with technical assistance, and to keep full documentation of planned conservation activity. While all US citizens owning or managing private lands are able to receive technical assistance, including assistance in developing a conservation plan, from USDA NRCS conservationists, free of charge, in practice most farmers who go through the conservation planning process are program applicants. An overview of technical and financial assistance for US hectares planted with corn in 2010, as surveyed in ARMS (published online 23 April 2015) is presented in Table A2. The participation in these activities is also separated into owned *versus* rented hectares (Table A2). In both cases, but more noticeably for rented land (10%), there is a lower level of participation (average overall 13%). For CSP, an even smaller contribution occurs, with 3.8% of hectares receiving payments (Table A2). Twenty-one percent of the area planted with corn has a written conservation plan, while less than 10% of these hectares have a nutrient management plan (NMP) (Table A2) [60]. However, in the US, NMPs are mandatory for manure management (and regulated by each state) but can also be used for synthetic fertilizer as part of conservation management. If the reported number is for mandatory (manure) NMPs only, it could explain the low number.

In 2010, although EQIP was obligated to spend $839 M to treat 5.2 M hectares of cropland nationally, it represents only a small proportion of the cropland area [60,61]. Ribaudo [60] assessed the distribution of EQIP contract obligations by resource concern (2008–2010). He found that: 13% of funding was spent on soil erosion prevention; 7% was spent on soil condition; and 36% of funding was spent on water quality and quantity, with the remainder spent on other resources, such as air quality and wildlife habitat. No similar data were readily available for CSP. In 2010, CSP was obligated to spend $320 M and enrolled 10 M hectares of land. This is almost double the 5.2 M hectare enrollment target set by the 2008 Farm Bill [60].

With only a small proportion of Australian and US farmers participating in soil testing and farm/conservation planning, it seems that the majority of farmers are soil-information poor, or at least they are not informed as to the condition of their soil by soil testing or by farm planning to highlight those soils at risk of land degradation. They could be relying on observational indicators for concerns such as soil erosion, but few studies have examined farmer use of these indicators [62,63,64]. Hence, a majority of land managers may be unaware of the deteriorating nature of their soil beyond the most basic of indicators. Without specific soil property information, farmers may similarly be unaware of what to do to impede the loss of soil condition.

4. NATURE OF SOIL TESTING AND FARMER MOTIVATION TO PARTICIPATE

The previous section revealed limitations of national-level reporting on farmers' soil testing practice which, in this section, has led to our reliance on regional studies or direct communication with those involved in soil testing, e.g., scientists, soil testing facilities and government employees. For instance, we contacted the International Plant Nutrition Institute in the US (IPNI) and the Australasian Soil and Plant Analysis Council (ASPAC) in AU for information on the volume of soil tests and the type of soil measures being undertaken. This section examines the nature of soil testing, motivation for its use (both perceived and actual), frequency and intensity of undertaking, as well as the types of soil properties most likely to be measured. Lastly, it focuses on the growing farmer interest in soil measures which examine soil health, and in particular, soil biological activity.

There is much assumed about farmers' soil testing practices, especially their reasons for soil testing, which we discovered is not supported by national-level data which do not ask about farmers' motivation to undertake a soil test. However, as a 2012 report [43] on agricultural resource and environmental indicators demonstrates, the perceived reasons why farmers test soil continue to be promulgated as: "*US farmers have used BMPs (such as soil testing for nitrogen and timing nitrogen application) to reduce fertilizer costs, increase profitability, protect the environment and conserve natural resources.*" [43]. It also seems that scientists and industry organizations differ in their opinion on farmers' interest in soil testing and their reasons for doing so. When these two groups analyzed the same set of data on farmers' participation in soil testing, there were two contrasting views published. One paper stated, "*It is apparent that soil testing is a relatively common practice in Australia, both in terms of number of farms and cropping area.*" [36], while another publication said: "*... soil testing is used sparingly as a tool to underpin fertiliser decisions. Some grain producers typically conduct soil tests at least once every 3 years on a selection of individual fields, but it is broadly understood that many grain producers use soil testing rarely or not at all.*" [65]. Clearly, what one group (scientists) considers routine is quite different to another (industry bodies). The contrast in perception of soil understanding between scientists and farmers is well documented in Ingram *et al.* [66,67], and even though farmers may have the "know-how" they do not necessarily have the "know-why". They found advisors were critical of farmers' soil management, saying *"they did not know how to examine their soil to determine how much cultivation is required"*.

Not only are researchers confounded by how often soil testing and monitoring occurs, it appears that farmers are too. Part of the issue is that soil health monitoring at the farm level does not usually occur in a formal, documented manner, with the previous section showing the low level of farmer participation in farm planning (Table A1 and Table A2). A qualitative study of grain farmers in northwest NSW, AU, showed [64] farmers felt they were "continually monitoring their soil", yet only 16% had retained soil test records at the paddock level or knew of its current status [64]. In response to the direct, open-ended question "How do you monitor the soil?" most farmers mentioned soil testing (67%) and visual appraisal (52%) [64]. Crop yield and protein level were mentioned by less than a quarter of farmers interviewed [64]. A small qualitative study (*n*= 15) of grazing farmers in Billabong catchment, NSW, showed that while they relied strongly on soil testing to support their management decisions such as the amount of lime to apply, they only focused on a few results such as phosphorus, aluminum and soil pH even though the soil tests would have provided more data on their soil [68]. These examples demonstrate awareness of soil function beyond productivity that still falls short of the ability to manage soil sustainably.

Certainly, in retrospect, the target set in 1999 in the Commonwealth Discussion Paper on Sustainable Land Management [69] under "indicators for capacity building for improved natural resource management" seems overly ambitious, and given the nature of national-level reporting, impossible to determine:

By 2005 there should be a 75 per cent increase in the number of landholders and regional communities actively monitoring resource condition—for example, by soil testing and water and biodiversity monitoring—to guide their management practices.

This is especially true as national surveys in Australia are no longer asking farmers if they test soil and also in light of the often informal or observational nature of soil testing and monitoring [38].

Nevertheless, a push to mobilize farmer interest in soil health and understanding was reinvigorated in Australia by a funding program, which operated from 2005 to 2008, called Healthy Soils for Sustainable Farms. A study, conducted as part of that program in the Lachlan and Macquarie valleys of NSW, asked landholders to rate a number of factors according to importance for their land management decision-making. Organic matter and soil structure were rated as "highly important" by 89% of those surveyed (*n* = 132 in 2005 and *n* = 131 in 2008) [70,71]. Despite these ratings, approximately a quarter of all broadacre agricultural businesses indicated they were soil testing in 2010–2011, with 72% of those including tests for organic carbon or 18% of all broadacre agricultural businesses [72]. Unfortunately, ABS does not ask landholders whether they measure soil structure. A review for the National Soil Research, Development and Extension (RD & E) Strategy [51] confirmed the dearth of physical and biological soil testing. The review showed that the 531,000 AU soil samples analyzed in 2012 were 87% chemical properties (including soil pH, nutrients, electrical conductivity and soil organic carbon (SOC), although this also has a biological component). Physical and (non-SOC) biological soil testing constituted 12.5% and 0.5% of the samples, respectively. However, this does not rule out the possible use of observational indicators for physical and biological monitoring by farmers.

In the US, the International Plant Nutrition Institute (IPNI) has collated the laboratory results of approximately 75% of North American testing laboratories, and has published these about every five years since 1990 [46,73,74,75,76,77,78]. They have clearly stated that the value of such a report is "*in calling attention to broad nutrient needs, in motivating educational and action programs, and in reminding farmers and their advisors of the importance of a soil testing program to monitor soil nutrient status*" [46], but they were also careful to point out that it is not at the farm level but across a state. It also provides an insight into the volume of soil testing in North America, and the types of tests undertaken, which have clearly increased in volume, especially in the Corn Belt of the US, which comprises 68% of the 2009–2010 soil sample volume (Figure 4). In their 2010 report, they indicated that 4.4 M soil samples were tested for P, K, and pH, while testing for Mg and S was done in about 2.6 M soil samples [75]. Their reports predominately focus on P, K and soil pH.

We visited a private US soil testing laboratory, Ward Laboratories [80] in Kearney, Nebraska, in May 2015. It is a typical soil testing facility. This laboratory processed over 350,000 soil samples in 2014, mainly for the basic level of soil testing which includes: pH, macro- and micro-nutrients and salts. Although anecdotal, communication with the laboratory personnel indicated that soil testing is a process usually undertaken on a cycle of three to five years per field, or maybe annually if there is irrigated land. Similar to our AU findings, this US laboratory emphasized soil chemical characterization and provided only a limited understanding of soil biology, soil structure or soil texture. At the time, they were undertaking a major expansion of their facilities, which in itself is evidence of an increasing demand for soil information by the farming and ranching public. This expansion includes the addition of some biological assays (in 2014 about 1% of all annual samples) to mirror the interest many innovative farmers have in these tests.

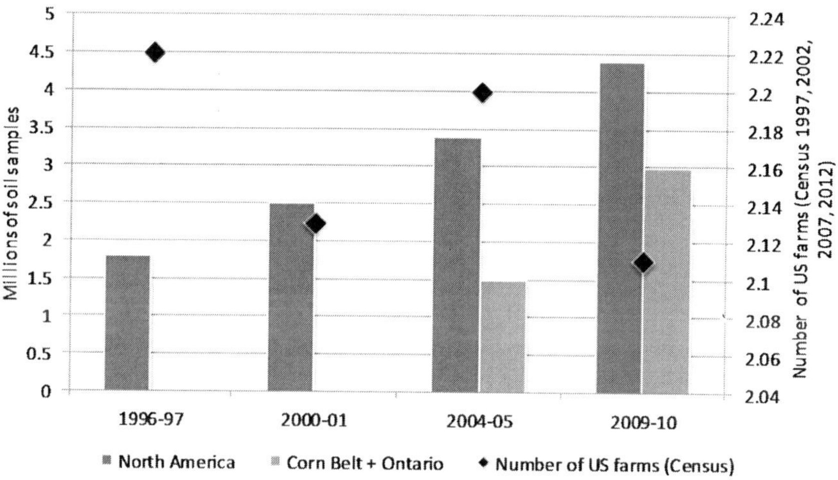

Figure 4. Volume of soil samples placed in North American soil testing facilities (over 14 years), and a number of US farms. Data sources: [45,46,73,75,78,79].

Similarly, innovative US farmers, who use various combinations of organic, no-till, rotation and cover crop conservation practices, were frustrated with the limited amount of information available to them via conventional soil testing. During the US farmer focus groups, these frustrations became apparent.

"It is astonishing to me after all the years I've been doing this that they're still only giving me a one page soil test. ... You need a more sophisticated tool than that."

—CA farmer

"I might as well be speaking to [crop advisors] *in Russian, about worrying about carbon levels or this or that, or C to N ratios, I mean they don't care."*

—CA farmer

"You get your samples and you take some of the recommendations, you put what they say and it doesn't really help you."

—CA farmer

The interest in soil biology is evident in many US conservation-related agencies. For example, soil biological health as a way of measuring soil performance has been recognized by the USDA NRCS in several ways. Training in soil quality or health has been available to employees since 1998. The CSP has included a soil quality component as part of eligibility since its inception. CSP rewards those who are already conserving resources and are willing to do more. The additional conservation activities are called "enhancements" [81]. Cornell's Soil Health Assessment [82] a service by which producers in the northeastern US can have chemical, physical and biological soil properties tested, is an available enhancement for CSP applicants in New York State. In 2015, another soil quality assessment, this one focusing on soil respiration and still under investigation by researchers, is currently available nationally as an enhancement (SQL15) [83]. Although farmers contractually agreeing to perform this assessment must collect samples and send them to participating laboratories, they are not required to act upon the information garnered by the tool. Many soil building practices are possible enhancements as well, including cover cropping, intercropping and controlled traffic. The number of nationally available soil quality enhancements has grown from seven in 2010 to 19 in 2015, mirroring the increased emphasis on the topic within the USDA NRCS agency.

During the focus groups with California farmers, participants communicated the need for monitoring information.

"I don't really know if I'm going in the right direction or the wrong direction on everything I'm doing. ... it takes a long time to impact your soil."

—CA farmer

When asked about their preferences for soil quality information, these innovative farmers wanted both an overall assessment of soil health over time and individual indicator information [28]. They indicated that this approach offered them adaptive management options as well as the ability to address specific resource concerns. Another US farmer recently shared with us how he used the soil information and insights he gathered to inform his practices. Much of his soil testing was to determine soil health. He was particularly interested in how active the soil biota were in mineralising N, and he used this information to consider his fertiliser practices. His particular interest in collecting and examining soil information shows that there is probably a small niche of farmers motivated to undertake such data collection and interpretation [84].

A regional survey of Australian grain growers in 1994–1995 shed some light on farmer motivations for soil testing. The proportion of respondents using soil testing to detect the change in soil condition was 11%. This was half of the 22% of responding grain growers that indicated an awareness of declining soil fertility in the previous five years [85]. The *GRDC Farm Practices Baseline Report 2010* [36] also had questions in their survey that examined the motivation behind soil testing. They found that it was largely to determine fertilizer requirements, not for soil monitoring, but the question was closed so no other options could be offered and confounded by including "other predictive tools" (Q158 *"And of your arable land, what percentage of it, has its fertiliser use based on soil tests or other predictive tools*?") [36]. Nevertheless, it seems unlikely that the majority of farmers' current use of soil testing conforms to the strictest notion of soil monitoring, *i.e.*, the nature and level of soil sampling is not conducted in a way that would enable determination of soil change [86]. Admittedly, we cannot examine the commitment to soil testing as we were unable to undertake cohort analysis of the national data so we cannot be sure if the farmers who indicate they are soil testing, in the various surveys, are undertaking re-measurement on the same areas.

5. EVIDENCE OF FARMERS USING SOIL INFORMATION AT THE FARM LEVEL

The rationale often advanced for soil testing is to provide useful input into farmers' decision-making about land management or problem identification. We would also add there needs to be a clear decision-making process, in which, if certain soil properties exceed or are below thresholds, specific actions are required to bring the soil condition back to acceptable levels. Even so, this type of information, while valuable, may not be enough for farmers to know what management changes are required to restore soil health. For instance, to indicate the positive outcomes of soil health management information, comparing soil measures, such as soil carbon or soil compaction, in the same or similar soils under various management systems and/or for records to be kept over extended time periods, *i.e.*, monitoring, is necessary.

5.1. Can Soil Testing Be Linked to Decision-Making?

In an attempt to link soil testing to decisions about altering specific land management practices (e.g., applying fertilizer), we compared fertilizer application rates and methods used to determine the amount required, using ABS REAACS survey data from 2007 to 2013 (Figure 5). Results show that 38% of those farmers who apply fertilizer are doing so on the basis of soil testing (farmers could select more than one method). Of course, that means 62% of farmers who apply fertilizer do not use soil testing guidance. Further, soil testing by farmers as a decision-making tool is trending downwards over time (Figure 5).

The ARMS Corn Production Practices and Costs Report, for 2005 and 2010, contain data on US fertilizer application rates and decision-making (Figure 6). At the field level, despite the wide use of fertilizer and relatively affordable soil testing prices, decisions regarding amounts of fertilizer were not guided by soil testing (Figure 6). In fact, despite almost 100% of corn farmers using N fertilizer, almost 70% of farmers were most likely to apply N based on "routine practice" (Figure 6).

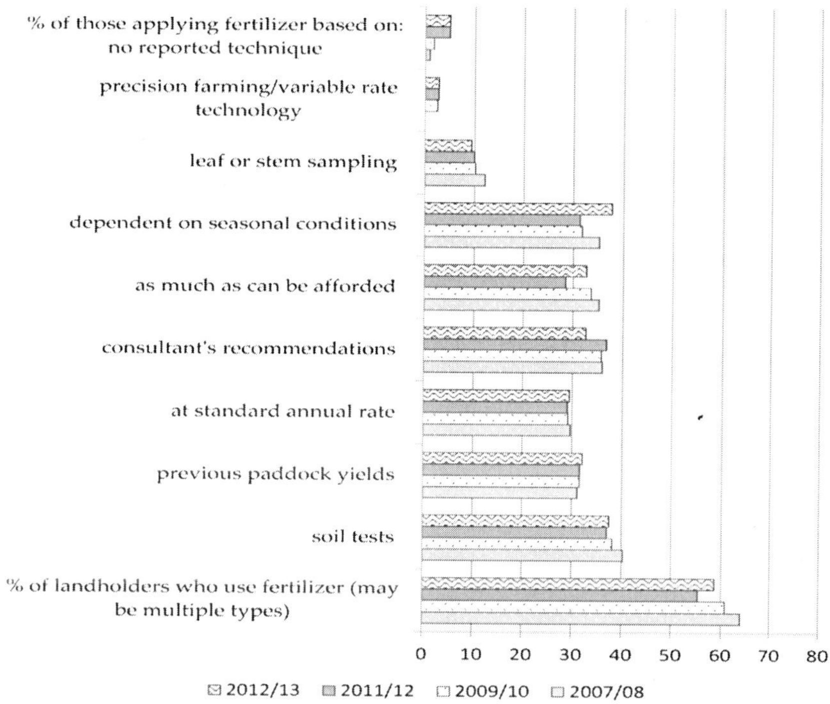

Figure 5. Summary of landholder self-reported decision methods for fertilizer application rate and the proportion of landholder-reported fertilizer use in AU (Data sources: [30,31,32,33]).

5.2. Links Between Soil Information and Adoption of Soil-Building Practices for Soil Health Management

The role of soil information, through soil testing or farm planning, is to provide the necessary impetus to change, if soil health is compromised, by adopting a practice that will counteract the negative developments, e.g., loss of organic C, nutrient depletion and increasing acidity. There are a number of widely accepted land management practices which have established links to soil health improvements. For example, no-till and stubble retention protect the soil from loss of soil condition by wind and water erosion, conserve soil moisture, and prevent loss of soil carbon [17]. In AU, it appears a number of these practices have been adopted by broadacre cropping farmers. For example, approximately 78% of crop and/or pasture land was under minimum or zero-tillage in 2010–2011, which is an increase of 55% from the 1995–1996 levels of 20% [88]. Interestingly, much of this adoption occurred without direct and localized evidence of soil degradation at the farm-scale. So even if soil health is a motivational factor for certain practice change, it does not seem necessary for farmers to collect soil test information in order to determine if the practice is required nor, once implemented, to determine if the soil health benefits are realized. Again, there is no information available about whether farmers are using alternative means of gathering information, such as observational data. However, if this is the case, it is unlikely that it is documented in any formal way by the farmers themselves either.

The benefits of improving soil health may prove to be persuasive for practice change, but a range of financial, environmental, social and personal motivators also play a role in achieving it [89]. Cropping practices can be at least partially motivated by environmental factors. Ecker *et al.* [57] states that "improvement in soil quality" was an important part of farmers' environmental motivation, along with "reduce[d] soil loss" and "reduce[d] water run-off runoff." These environmental concerns, however, ranked second behind financial motivators. Nevertheless, the survey instrument used pre-determined, pre-categorized responses to close-ended (yes or no) questions. This approach can be misleading: it may not necessarily reflect the farmer's full intentions. In a smaller qualitative study by CAAANZ, reported in Kancans *et al.* [85], the researchers were able to more deeply assess some environmental motivations around soil condition, with farmers' statements, such as:

"Soil to me is the key to the whole system. I wanted to change the system to improve the soil"

—AU farmer

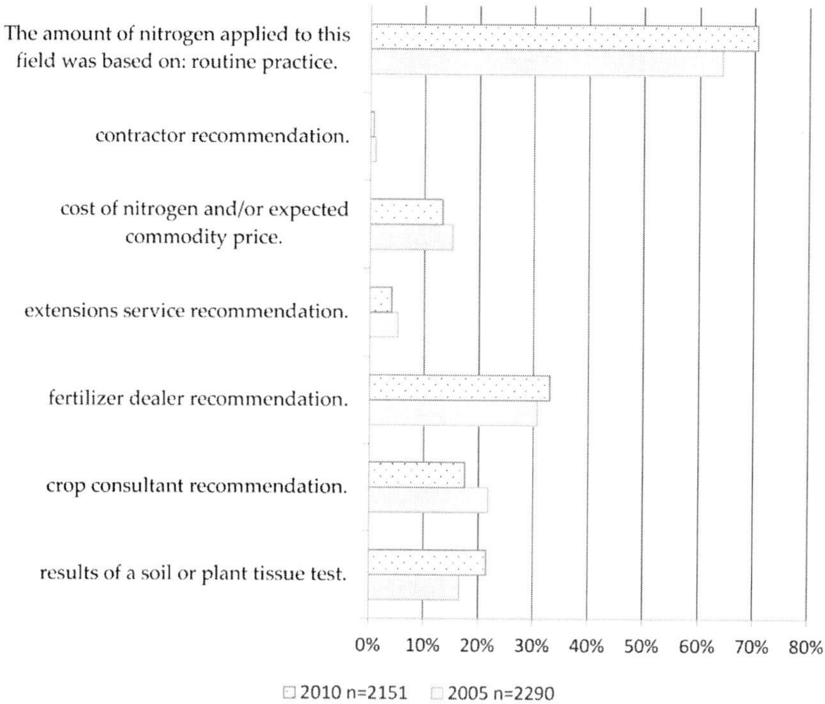

Figure 6. Decision-support tools used by US corn growers to determine nitrogen application rates, as reported in ARMS surveys in 2005 and 2010 (Data Source: [87]).

A number of the study participants commented on improvements in soil health, which were largely observational, including the re-appearance of earthworms, which were not necessarily informed by soil test results [90]. A longitudinal study of farmers' adoption of no-till with stubble retention found the following adoption rate influences, in priority order: lower fuel and labor costs (at seeding); soil conservation, *i.e.*, less soil erosion; and soil moisture conservation [91,92]. In California, US, farmers' adoption of no-till or reduced till seemed to have similar motivations: the adoption rate increased in proportion with rising fuel costs, while stricter air quality regulations had a much smaller effect [93]. Interestingly, despite soil conservation being a primary driver of no-till adoption, there is not a lot of quantitative evidence, as collected by farmers, to demonstrate that soil erosion and other soil health concerns have lessened under no-till, despite the scientific research that indicates it does [1]. Nevertheless, farmers do speak of reduced fuel and labour costs as a primary result of fewer field passes [91]. The reason most farmers gave for non-adoption of no-till in one study was "lack of observed benefits" [91].

Kelly *et al.* [68] noted that farmers in their study believed healthy soil to be more a "means to an end" than an "end in itself"; rather, their goal is a soil capable of producing sustained yields. This sentiment was echoed in the forward of a prototype for a Soil Health Management Plan which stated: *"Soil health is not an end in itself. Soil, water, sunlight and air are the primary natural resources for agricultural production"* [94]. It was proposed to incorporate the Soil Health Management Plan into farm planning in Victoria, AU, which, as we explained earlier, was at the most 30% of farm businesses that have a farm plan in Australia (Table A1). This concept of sustainability was also echoed by the US focus group farmers, who made statements such as:

"A lot of times it will be less expenses. And [healthy soils] will *be more manageable, soil structure is better... Your plants are healthier. You need less inputs. Your plants just grow better."*

–CA farmer

"...whatever you've done to that field or whatever condition it's in, it requires probably the least input and gives you the best results."

–CA farmer

In the US, we examined cover crop usage and the role of soil health and soil information usage in leveraging its adoption. There are numerous scientifically documented benefits of cover crops on soil health such as sequestering atmospheric carbon in plant tissue, providing habitat for soil biota, and stabilizing or enhancing soil aggregate strength [95]. However, while 82% to 94% of crops are grown in some sort of rotation, conservation crop rotations that incorporate cover crops remain uncommon [96]. Only about 3% to 7% of US farms use cover crops in rotations. Additionally, since these operations do not put all of their land into cover crops, only 1% of cropland acreage uses cover crops [96]. The USDA Conservation Effects Assessment Project (CEAP) estimates that from 2003–2006, cover crops were used on less than 1% of the Upper Mississippi Basin region, which drains approximately 49 M hectares, including large parts of the states of Illinois, Iowa, Minnesota, Missouri, and Wisconsin [97]. Singer [98] surveyed 875 farmers in Illinois, Indiana, Iowa, and Minnesota (in the US Corn Belt) on their cover crop use in the autumn of 2005, finding that 8% of the farmers planted cover crops that year, while 11% had used cover crops within the previous five years.

Given these low rates of adoption of cover crops, Wallander [96] suggests that *"further research could help identify the types of incentives that may be needed to promote wider adoption of cover crop rotations."* Likewise, a small group of 36 farmers, scientists, extension specialists, and policy experts convened by the National Wildlife Federation in June 2012 devised a "Roadmap" for addressing the barriers to cover crop adoption [61]. Their report identified research priorities to increase farmer adoption. The study of the *"environmental impact of cover crops was 5th on the priority list, after benefit-cost analysis, establishment options, impact on commodity crop yield and species selection and combination, in order of priority."* This may suggest the benefits are relatively well recognized compared with how to implement the practice. Nevertheless, none of the publications reviewing research needs suggested that being better informed about soil condition, e.g., by soil testing, could encourage more people to adopt the practice by enabling documentation of soil health benefits.

To examine landholders' interest in cover crops and their motivation for planting them, the Conservation Technology Information Center (CTIC), along with funding from the USDA NIFA Sustainable Agriculture Research and Education program (SARE), published two cover crop surveys, in 2012–2013 [99] and 2014–2015 [100]. The 2012–2013 survey was administered to no-till field day participants throughout the Corn Belt and Great Plains regions. The 2014–2015 survey was administered online and was accessible to all US farmers. However, the researchers caution that both their participant groups are likely to be skewed toward farmers already interested in cover crops, due to the field day topics and locations of advertisements in no-till and cover cropping magazines. Out of 1200 respondents in 2014–2015, 84% had grown cover crops [100], which is in stark contrast to earlier state-wide surveys which show cover crop adoption at less than 12%. This bears out the caution from the researchers but also points to the types of growers most likely to adopt the practice, which could assist educators with whom and how to approach the topic. From

2010 to 2013, there was a 30% increase per annum in reported cover crop area sown for those surveyed cover crop users [99]. Both types of respondents, those growing cover crops and those who did not, ranked time and labor as their biggest concern with cover crops [99,100]. However, when cover crops growers were asked to list their top three benefits, the leading benefit was "increases in overall soil health" (22%), followed by "increases in organic matter" (21%), and "reduces soil erosion" (15%) [100], all three responses pertain to soil health. Despite soil health being cited as the key reason for adoption, it is unlikely many of these growers have documented evidence showing changes in soil health given the low participation in traditional soil testing (at about 30% of farmers [39]).

Interestingly, farmer participants in these surveys said agricultural retailers could help them assess soil change as well as help them use that information in their nutrient management plans. This is similar to the CA focus group theme that, while interested in soil health, they would rather pay consultants to do the soil testing for them. This makes sense given that there are fewer people managing greater land areas, as indicated by the decline in numbers of farms, especially in Australia which has declined by 40% (1981–2011) [101], and the US which has declined by 15% (1982–2012) [45].

Thirty percent of the cover crop survey respondents who were actively using cover crops reported receiving some level of cost-share assistance to do so [100]. Of those surveyed, only 9% said they would only plant cover crops with financial assistance [100]. The authors concluded that incentive payments were important to a minority of farmers, but most growers would continue the practice without any financial assistance [100]. Perhaps more importantly, economics ranked well below the numerous soil health impacts, such as improved soil health, SOM and reduced erosion, in farmers' lists of benefits for both users and non-users of cover crops [100].

5.3. Use of Soil Testing Data in Application of Lime

Soil pH plays an important role in regulating nutrient availability and influences the environment for soil biological activity. Therefore, liming acidic soils is a practice of considerable value to soil health. The AU National Land and Water Resources Audit (NLWRA) estimated soil affected by soil acidity was 50 M hectares of surface soil, with 11 M hectares of highly acidic soils requiring 2 t/ha/yr to increase or maintain a pH_{Ca} of 4.8 in 2001 [102]. Current liming rates are well under the levels required to arrest the problem [103]. Therefore, we examined the use of lime as it relates to soil testing in Australia.

In the REAACS ABS survey from 2007–2013, questions were asked about testing for soil pH, and the management of soil acidity. Only 9% of the affected area (1.8 M ha of agricultural land) was treated with lime across Australia [33]. Over half of the lime applied in AU during 2012–2013 was in Western Australia. Even though annual applications of lime have risen sharply over the last 10 years to just over 1.6 M t lime/year in 2014 sold (and presumed applied), the application of another 2.5 M t lime/year is presumed to be required to abate soil acidity in Western Australia. Notably, Gazey *et al.* [103] showed that when Western Australian farmers increased the frequency and intensity of soil testing, it led to increased applications of lime.

A relatively high proportion of those agricultural businesses reporting soil and land issues were aware of soil acidity and soil compaction, but few farmers tested soil physical properties (Figure 7). For those businesses reporting soil activities, a similar proportion were also liming or soil testing (Figure 7a,c). The 2006–2007 ABS survey of land management practices [104] showed lime applications (Figure 7c,d) performed by 30% of all agricultural businesses. Over these two sampling periods there appears to be an increase in lime application, although the testing of soil acidity has remained constant (Figure 7a,c). However, it is impossible assign attribution to increased problem awareness through soil testing and increased remediation activities from these data as no cohort analysis exists. When the actual proportion of those reporting testing soil pH and liming was calculated as a proportion of all agricultural businesses, results suggested awareness and actions overall were limited to about one quarter of landholders (Figure 7b,d) [104,105]. However, in more recent surveys (2012–2013), the application of lime and soil testing has seemingly declined, with only 9% of all agricultural businesses applying lime, while 16.7% of all agricultural businesses had tested soil pH in the same period [33]; however, this could be more related to the post-2007 surveys having a defined period to report on.

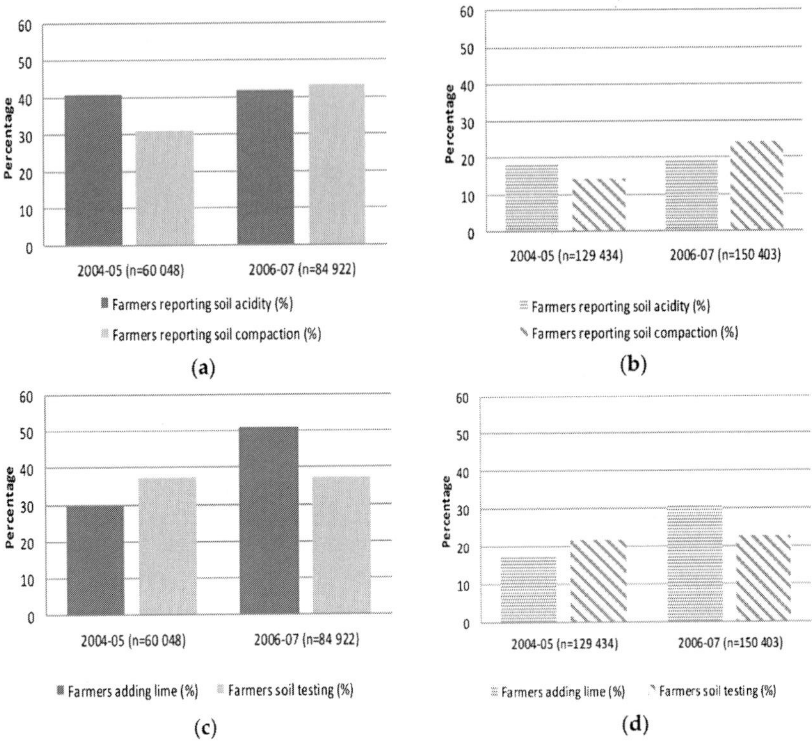

Figure 7. In Australia, changes in awareness of soil acidity and compaction issues*versus* lime application and soil testing, shown as a proportion of those reporting (**a**,**c**) and as a proportion of the total number of agricultural businesses (**b**,**d**). Data sources: [104,105].

6. LIMITATIONS TO WIDESPREAD USE OF SOIL INFORMATION FOR SOIL HEALTH MANAGEMENT

In Australia, one study found two impediments to a structured soil health management plan: high cost and insufficient one-on-one expert advice or assistance to support a change [71]. Work conducted by Bramley and Cook [106] suggested that trying to use soil testing results for two different purposes, crop production and soil resource assessment (e.g., soil health), was problematic due to differences in soil sampling intensity needs. Additionally, they state that bias towards those farmers who are more proactively managing their soil is not representative of soil condition. Their conclusion stated that *"if we get soil testing [suited] for the purposes of crop and fertiliser management it will ... become a more useful tool for soil resource assessment. This assumes ... also that soil testing becomes a more common practice for a wider cross section of producers."*

US focus group participants mentioned a number of barriers to the adoption of soil building practices. One theme was the issue of economic costs:

"I think the more and more farming becomes tougher economically, then everybody continues to look shorter and shorter term. ... So, sometimes that limits what you think might be the optimum, long-run approach."

—CA farmer

This echoes the concerns farmers had with cover crop expenses in the CTIC survey [100]. Another very prominent theme in the US focus groups was the uncertainty of land tenure:

"We have leases anywhere from one year to maybe five years. That's about as long as it gets. ... I know of several farmers in England and they have 30- and 40-year leases over there. ...And you have a different attitude and you can do different things."

—CA farmer

"Instead of a [grass] waterway, you surface cut that channel to the ditch or to the river and farm it. Because I'm paying cash rent on ground and by God I'm going to I'm going to something out of it."

—IL-IA farmer

Although it is well known that land tenure affects management decisions [107], this does not seem to be considered a soil health issue. In the US, farm land is approximately 60% owned and 40% rented. However, this percentage changes regionally. Some of the highest rental rates are found in the most intensively cropped regions, such as the Corn Belt and Central Valley, CA (which were the strategic locations of our focus groups), where rental rates are often greater than 50% [107]. This leaves farmers with low incentives to invest in the health of soils they may not be working next season.

It seems that those who do engage in soil testing and farm planning would do so regardless of the incentives. However, others will require incentives to see the environmental and economic imperative of gathering information to aid in sustainable management. Alternatively, farmers may need assistance from crop advisors or agricultural retailers to do testing or monitoring and provide management advice from that information. A recent convert to data collection, including soil testing, is a farmer in Iowa: "I was excited to join the Soil Health Partnership (in 2014) because of the emphasis on data collection," Smith said. "Documenting the benefits will help promote the practices." [108].

6.1. Specific Impediments to Soil Testing

Cary *et al.* [102] examined a number of sustainable practices based on Rogers' attributes [109] of a land management practice that facilitate adoption, including: relative advantage, compatibility, complexity, "trialability" and observability. Regular (Note: regular in most cases is not defined and neither is the time period it applies to, and hence it could be interpreted differently by landholders) soil testing scored well for risk and complexity, but rated poorly under relative advantage, "trialability" and observability. Soil and plant testing for fertilizer requirements scored poorly under complexity, and was similar under the remaining categories when compared with regular soil testing. A US study also stated that despite the advances in soil laboratory procedures and analysis, fertilizer response calibrations, and computer use for data processing and reporting, soil testing still suffered from poor credibility due to the wide variety in recommendations [110].

Although we made the assumption that soil testing was a reasonable proxy indicator of soil health management, during the course of this investigation, barriers to widespread use of soil testing for this purpose became apparent. The application of soil testing data as a sensitive and reliable tool for detecting soil change is met by a number of obstacles, which are:

(1) Soil sampling intensity:

- Site-specific issues are missed with bulked or zonal sampling [106].
- Soil spatial variability is too great and sampling intensity too low to give accurate information for many properties [46,111,112].
- Standard soil sampling depth (0–20 cm) by farmers can be too deep to identify changes in soil surface properties, or equally too shallow to detect sub-soil changes.

(2) Soil sampling frequency:

- Soil testing may be too infrequent [112] for adaptive management uses.
- The trajectory of change may be difficult to plot with only a few sampling points over time
- Records of soil tests are not retained to allow comparisons to be made over time [64].
- Soil testing may be conducted every three to five years but not necessarily in the same paddock.
- The slow rate of soil change in certain soil properties, coupled with the lack of outward visibility of some types of soil health loss, such as declining soil pH, can be misleading [11].
- Monitoring soil health is difficult without a baseline taken before management changes were implemented.

(3) Lack of holistic focus:

- Conventional soil testing does not monitor the changing nature of soil structure or soil architecture in terms of water availability and habitat for soil biota, a major source of soil health improvement [113].

- Soil testing is very much focused on soil chemical properties and, indeed, a landholders' focus is often only on a few key measures such as macronutrients (P, and N), and soil pH [74,75,78].

- The metric of soil health performance is often related to the productivity function of soil by relating it to yield, rather than improving other functions such as soil biological activity and the flow-on consequences of that particular decision [113,114].

A few ideas worth exploring are the blending of observational indicators with conventional soil testing data, standardizing of observational indicators, and developing closer working relationships between farmers and soil professionals. In the 1990s, there were a series of developments in the US [62,115,116], many sponsored by USDA-NRCS,that sought to assist the farmer or crop advisor in gathering their own soil information. Guidelines for developing a soil health scorecard for local areas were among the products developed by the NRCS-Soil Quality Institute [117]. These products and a number of state Soil Health Cards are still available online [116]. Another tool developed in the 1990s by ARS and NRCS was the Soil Quality Test Kit [115,118], which was intended for use by farmers and conservationists. Such tools were advised to be used in conjunction with *"routine soil test information that currently focuses almost exclusively on soil chemical or fertility parameters."* [119]. In light of the current review, revisiting these tools, with the aim of training consultants and educators, may be warranted.

6.2. Missed Opportunities for National Surveys

Issues that we encountered during this review offered insights into the gaps in national-level reporting of soil testing in terms of understanding farmers' participation (both frequency and intensity), and the nature of soil testing and motivation behind it. The data gaps had implications for the type of conclusions that could be drawn from this data. One recurring issue concerned difficulties in attribution, either between or within surveys, especially examining links between soil testing and farmers' use of soil testing data in terms of their soil health management behaviors over time. In AU and the US, there was only one question which asked about soil testing in relation to fertilizer application. However, even with this question we do not know if the sample population from each survey period was the same or different. While specific concerns were discussed above, overall they can be summarized as: closed questions with built-in rationales for soil testing, unknown or unreported population size, no cohort analysis over time, rarely any examination of the relationship between soil testing practice and other soil building practices, changes over time in sampling areas, intensity and survey questions, unrestricted*versus* time-bound questions which made questions incomparable over time, surveys being conducted every five years and only asking about the previous or current year, inconsistent reporting due to a shift in reporting language from the original question intent, and limited reporting of findings on soil testing, even when data had been collected. Also, for the purposes of comparisons, reporting units were different between AU and US. ABS uses the actual number of businesses, extrapolated from a sample population (about one-fourth of all agricultural businesses), while in the US, data is reported as a percentage of farmed area, and is extrapolated from a smaller sample size of farmers' fields. More could be gleaned from these surveys if motivations for soil testing, and frequency or intensity of soil testing were addressed. We further suggest that questions about motivations for practice change be included to help educators with designing assistance. A few relatively small changes could potentially provide much-needed information about farmer information use.

7. IMPROVING FARMERS CONNECTION TO SOIL INFORMATION

7.1. Information Availability for Farmers

For soils in particular, access to information of local relevance is complicated by a disconnection between data sets and user location. The sampling effort in AU is greater in coastal zones, resulting in patchiness in data coverage and quality; it is also limited to the decade around 1990–2000, when most of the sampling effort was made [51]. The US has much better coverage of soil data, with most farmland covered by 1:12,000- or 1:24,000-scale soil maps, but this data has been accumulated over many decades of soil surveys, with no reported information on management at the time of sampling. Further, these data are on inherent, taxonomic soil properties, not the impacts of management on soil health.

In 2014, the Australian Soil RD & E Strategy [51] called for a national database of soil information to support farmers' management decision-making. The proposed model for soil information delivery would have national-level coordination, with delivery to various audiences, including farmers, from a central online repository. One option under consideration is to include soil testing data collected by private enterprise as part of the database, if data is available for the queried soil type and land management practice. In Australia, the current soil data held in public institutions is described as "patchy, incomplete and out-of-date" by one researcher [120].

This national database approach is similar to the USDA NRCS Web Soil Survey [121] for the delivery of US soil taxonomic, inherent property, and capability information. Also underway, as part of a joint USDA NRCS and USDA Agricultural Research Service project, is the development of a dynamic soil properties database to be developed by mining the primary literature for the effects of agronomic management on soil quality indicators. The website portal will allow users to access soil change information, using extrapolation techniques of the data to query locations that consider soil type and climate factors. The database will also be used to inform modeling efforts and perform meta-analyses of management effects. Ultimately, it seems that both countries are moving in the same direction to provide landholders with soil change information. However, neither country has achieved this goal at this time.

7.2. Connecting Farmers to Soil Information Sources

Increasing farmers' access to and use of soil information means engaging farmers more directly with existing soil information sources. Familiar traditional extension activities are still major information sources for agricultural land managers with field days and training courses as the key sources of information on sustainable land management practices [59], e.g., CTIC no-till field days in the central US. Of course, online mediums also offer opportunities for people to access expertise when they need it and when they are ready. Sixty-seven percent of US farms had Internet access in 2013, with 70% of farms having access to a computer. Farms using computers for their farm business increased to 40% in 2013 compared to

37% in 2011 [122]. E-extension [123,124] is being trialled in various topic areas such as crop nutrition and plant protection in AU, and US and in organic agriculture in the US, with some success in providing timely, useful information to the intended audience as well as sharing materials among educators and dealing with the loss of expert knowledge on the ground [125]. The US target audience is the community of practice, including farmers and educators.

Despite the increasing volume of information available online, the ability to engage with it is limited, due to its often static nature and frequently non-engaging presentation. Improved information flow (both in relevance and accuracy) to landholders is difficult when soil information is archived or in some cases "lost" (not updated, due to program completion or funding cuts, e.g., Land and Water Australia (1990–2009) [126] and National Land and Water Resources Audit (1997–2008) [127]. In Australia, information available online is increasingly stripped back or de-contextualized so it is only presented at regional or state scale, rather than at a useful scale for a land manager (e.g., Australian Bureau of Statistics data on Land Management and Farming in Australia [26]). Similarly, reporting may be only online with no downloadable version, while older publications (and, with them, the accompanying institutional memory) can be lost entirely. In the US, the use of digital archives is common, but potential audiences may not know to request archival information.

Increasingly, social media platforms such as Twitter are used by organisations to promote their activities (e.g., FAOKnowledge [128]). The NRCS USDA soil health web portal has a wealth of information, short videos and activities in state areas [129], but the general message is about raising awareness on the topic. Digital media are particularly useful for raising awareness of issues, to catch viewer attention and to stimulate interest, but online media are less powerful for practice change and discussion (e.g., Twitter is used by USDA NRCS to highlight their work and show how landholders are involved and can get involved in programs [130]). We must remember that information provision is not the same as creating a dynamic learning environment.

Both the ABS and USDA NASS databases on soil and land management practices can be interrogated. Unfortunately, the scale is not applicable to the typical land manager. Furthermore, available data does not include soil change information. The ABS data does allow for graphing functions, but lacks visual appeal, and it is static and is onerous to manipulate. One alternative to these large databases is the use of applications (apps) for mobile devices, such as SoilMapp™, developed by CSIRO in Australia, and SoilWeb, developed by the University of California Davis [131]. Both apps use GPS (geographically located) and are relatively easy to navigate. Both sets of data were collected over many decades. SoilMapp™ (CSIRO, Canberra, Australia) has been in the public domain for nearly three years and has been downloaded some 3600 times, but the audience is unclear, as are the developers' intentions for its use. SoilWeb uses NRCS US soil data on inherent properties, which are critical for interpretation but do not tell the story of soil change at the human time scale. These databases are also great education tools on soil types and their characteristics. However, by themselves, their value is limited by the lack of soil change information and other predictive capabilities.

At present there do not seem to be many online platforms that offer an interactive way of examining soil data and interpreting it for various users, with some small exceptions [132]. The soil quality website does allow farmers to compare their own soil testing data to data held in the website. The output is a graph of data collected for that locality, which shows, through traffic-light indicators, unacceptable and acceptable levels of the soil property in question, with variation according to broad soil texture classes. The website stipulates that for some soil properties, target values are known, and one-off sampling can be used to determine if a soil meets the target value. However, for other soil properties, target values are not defined and subsequent monitoring over time is required so that trends can be established. In the US, the National Soil Survey Center (NSSC USDA NRCS) offers a number of online soil portals that can be used, around soil survey and soil health, mostly to produce soil maps and tabulate soil data. These include Web Soil Survey [121], Soil Data Access, and Dynamic Soil Properties, which are all built from the same soil database, with over 2.5 M data records, called the National Soil Information System, (NASIS). Web Soil Survey (WSS) provides both soil maps and the soil data explorer provides access to soil data for your "area of interest" and determines the suitability of the soils for a particular use. Web Soil Survey has soil maps and data available online for more than 95% of US counties. Web Soil Survey is operated by the USDA NSSC and has 152,000 unique users each month and produces 493,000 customized soil reports per month [133].

As we have learned from several of the studies above, direct human interaction with face-to-face communication is still needed. Unfortunately, a capacity constraint for both knowledge generation and exchange is the aging nature of the workforce, both in agriculture and the service industries of education and extension [51]. Fortunately, the US Farm Bill targets beginning farmers in many of their assistance programs. Nevertheless, US farmers are aging, and in the last census of 2012, the average age was 58.3 years, with 33% of US farmers aged 65 years and over [45]. Australia's farmers also tend to be considerably older than other workers. In 2011, almost a quarter (23%) of Australian farmers were aged 65 years or over, compared with just 3% of people in other occupations [101]. The aging of the farming community in both countries also has implications for how knowledge is presented, as many have yet to widely adopt digital technologies.

7.3. Developing Social Networks for Knowledge Exchange

With an aging and smaller workforce comes the loss of experiential and expert knowledge which could be addressed in part with succession planning and the use of mentoring, but to date there seems to be little consideration of how to train the next generation of soil conservationists. Mentoring in universities, government and farming communities is a worthwhile endeavour to stem the loss of expertise and local experience. Alternately, using existing or developing knowledge networks (such as those that currently exist in NSW, Australia, with retired soil scientists) or remotely accessed nation-wide databases may in some way address these concerns. Hunt *et al.* [134] identified a positive feedback loop between reduced government involvement and declining employment opportunities, with fewer graduates and undergraduates studying those fields, further reducing the number of skilled professionals and tertiary resources for the study and practice of sustainable agriculture. A recent evaluation of the Australian Soil RD & E staff re-enforced the limited investment in soil knowledge exchange, showing 12.5% of its workers in extension and 6% in teaching, with the majority of staff in research (39%) and postgraduate training (26%) [51]. At the same time, there also seems to be an increase in farmers undertaking formal post-secondary school qualifications. Over the three decades to 2011, for instance, the proportion of Australian farmers with post-secondary qualifications more than doubled, from 15% to 38% [101]. The proportion of farmers with a certificate-level qualification doubled over this period, while the proportion with a bachelor's degree or above increased six-fold.

Making the connection between scientists and land managers requires great effort for developing social networks. While innovative farmers and enterprising scientists interested in outreach will pair up to do on-farm research and have field day discussions and even join e-extension communities of interest, beyond these subgroups communication is all too infrequent in our experience. Broadacre and grazing farmers in the Border Rivers-Gwydir (BRG) CMA region were found to be more likely to seek out other farmers for information on natural resource management rather than go to the CMA or to Landcare (Figure 8). They cite the strength of their ties as a factor [135]. The more worrying trend is the decline in these regional-level organisations as sources of information, especially Landcare (Figure 8). At its peak in 2002, Landcare had 40% of landholders involved, and in the last survey conducted in 2012, membership had slipped to just over 20% of Australian farmers. Soils-focused groups such as Soils for Life [136] and SoilCare (NSW) [137] exist because of farmer interest in managing soils better and a desire to share experiences and knowledge. These groups have developed largely outside the major RD & E structures, isolated from existing knowledge sources with few communication links to recognised research and development.

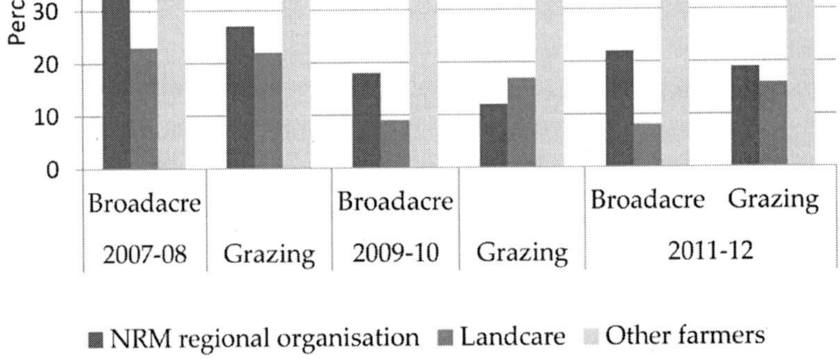

Figure 8. The percentage of farm businesses in the Border Rivers-Gwydir region, NSW, AU, that source natural resource management information from regional organizations, Landcare or other farmers according to main enterprise, divided by production type, for three time periods. Source: [30,31,32].

In the US, farmer groups are not subsidized by the government. However, US farmers have formed interest groups and have regular field days on conservation management topics, such as no tillage and organic agriculture (National Corn Growers Association Soil Health Partnership [138]). US farmer interest often dictates which information sources and networks have the most influence. For users of cover crops, the most relied-on sources of information (with equal importance) are: farm media, the USDA or local conservation district representative, and university extension educator, while for non-users the greatest influence for advice on cover crops was that from neighboring farmers, followed closely by the USDA or local conservation district representative and agricultural retailer [100].

8. CONCLUSIONS

For positive changes in soil health and agroecosystem sustainability to occur, the farmers are the most important people of influence, because ultimately it is their actions that will determine the state of the soil [66]. Researchers, educators and policy-makers must address the best "means to an end" by which they can achieve this. We argue that the link between the use of soil information, here indicated by the use of soil testing and farm plans to monitor soil health and improve soil management, and, ultimately, ecosystem sustainability, has yet to be proven, but essential to establish. Hence, the way forward must include a better understanding of how farmers currently gather and use soil information, by addressing the limitations of current national data sets.

Soil testing, as currently available to farmers, is necessary to document soil health change or improvement, but due to its low application (frequency and intensity), along with known limitations, is probably insufficient for this purpose. Even so, despite its flaws, the type of soil information delivered by soil testing, focusing on nutrients and soil pH, is underutilized, engaging less than 30% of farmers, at any one time, in both countries,; at least appears that way from the national-level reporting of soil testing practice. Anecdotally, we can show there is a small core of farmers interested in gathering soil information and applying it to managing soil health, such as the innovative farmers who participated in the US focus groups. Less evident is the commitment to soil testing over time by farmers and how it is used by farmers in their farming practice. Despite the recurrent language, in policy and other documents, that suggests farmers' monitoring of soil health is necessary to guide decision-making and land management practices [51], the reality of their practice, as shown in this paper, has been difficult to establish.

The US and AU databases fell very short of providing information on farmer motivations for soil testing. Even when questions about motivation were asked, the answers to those regional survey questions were not reported [97,139]. While this information is not the primary goal of these surveys, a few relatively minor changes, made in consultation with agricultural educators, could provide the much-needed information. Additionally, we found sharply contrasting conclusions made about the same set of data [36,65], as well as claims that appeared to be unsubstantiated by the evidence presented. There is little wonder that the understanding of what farmers do with soil information and why is rather limited.

Fortunately, there were some regional studies, in the US and AU, to draw upon and compare with the national survey results. As a proxy for soil health assessment, we maintain that soil testing was a reasonable "gateway indicator". However, there are many well-documented concerns on soil testing and its reliability [21,110,112], which may be one reason its use seems low. One study reported that farmers felt they continually monitored soil health, yet formal documentation of soil testing results was minimal, and instead they often relied on visual appraisals (52% of AU farmer respondents reported) [64]. A greater emphasis on observational indicators may gain strong acceptance from farmers. Participants in a US survey reported wanting assistance from crop advisors or agricultural retailers to do testing or monitoring and provide management advice from that information [62]. It seems clear that those farmers who perform soil testing require clear "how to" guidance. Those who do not use soil tests may prefer to pay consultants to perform testing for them.

We found obvious growth in soil building practices, such as no-till in the US and AU and cover crops in the US [91,92,99,100]. Improving soil health is seemingly a persuasive motivator for those farmers currently using these practices, regardless of policy incentives and farm-level information on soil health benefits. However, for those farmers yet to adopt such practices it seems they will require incentives to undertake soil building practices, as well as understanding the environmental and economic imperative of gathering information on their soil health. For instance, one US study concluded that incentive payments were very important to some farmers, but found that when environmental benefits were apparent, they continued the practice without subsidy [140]. However, for continued growth in soil building practices by farmers, the benefits must be observable, whether by ocular estimate or through laboratory tests.

The adoption of sustainable natural resource management practices relies on a number of social, political, economic and institutional factors to intervene and influence landholders' motivation and capacity to change [89,102]. Identifying land degradation or loss of soil condition, through soil information, would seem a critical first step to land management change, but as shown in the adoption of no-till, it has occurred without such information at the farm scale. Farmers seemingly recognize the long-term commitment needed to build soil health, but our understanding of their interest in soil health status is limited due to the nature of national-level data collection on soil testing practice. Policy-makers can address this knowledge vacuum by supporting improved national-level reporting and data analysis. Educators and other agricultural professions can provide guidance and consultation to farmers on how to assess soil health changes in response to management. Ultimately, it is farmers who must make the needed practice changes. To make and maintain management improvements, farmers want to be confident that the benefits to their soil are real. Soil assessment and monitoring can benefit us all if undertaken with clear guidelines and within supportive structures.

ACKNOWLEDGMENTS

No funding was used in the undertaking of this work. Thanks go to Jodie Mewett from ABARES for provision of data cubes. Thanks also to Rachel Brown from USDA ERS NASS for supplying data from ARMS surveys for Figure 6. Thanks to the various scientists and landholders who freely shared their views on soil testing and its uses.

AUTHOR CONTRIBUTIONS

Lisa Lobry de Bruyn did most of research for and writing of this paper. Susan Andrews contributed discussions, the US focus groups results, major editorial revisions and organization and some writing.

REFERENCES

1. Pimentel, D.; Harvey, C.; Resosudarmo, P.; Sinclair, K.; Kurz, D.; McNair, M.; Crist, S.; Shpritz, L.; Fitton, L.; Saffouri, R.; *et al.* Environmental and economic costs of soil erosion and conservation benefits. *Science* **1995**, *267*, 1117–1123.

2. Food and Agricultural Organization of the United Nations (FAO). *Revised World Soil Charter*; FAO: Rome, Italy, 2015; pp. 1–7.

3. Dobbs, T.L.; Pretty, J.N. Agri-environmental stewardship schemes and multifunctionality. *Appl. Econ. Perspect. Policy***2004**, *26*, 220–237.

4. Karlen, D.L.; Andrews, S.S.; Doran, J.W. Soil quality: Current concepts and applications. In *Advances in Agronomy*; Academic Press: Waltham, MA, USA, 2001; Volume 74, pp. 1–40.

5. Pretty, J.; Brett, C.; Gee, D.; Hine, R.; Mason, C.; Morison, J.; Rayment, M.; van der Bijl, G.; Dobbs, T. Policy challenges and priorities for internalizing the externalities of modern agriculture. *J. Environ. Plann. Manag.* **2001**, *44*, 263–283.

6. Soil and Water Conservation Society (SWCS). *Beyond T: Guiding Sustainable Soil Management*; SWCS: Ankeny, IA, USA, 2008; p. 18.

7. Committee on Long-Range Soil and Water Conservation; Board on Agriculture; National Research Council. *Soil and Water Quality: An Agenda for Agriculture*; National Academy Press: Washington, DC, USA, 1993; p. 519.

8. Larson, W.E.; Pierce, F.J. Conservation and the enhancement of soil quality. In *Evaluation for Sustainable Land Management in the Developing World*; International Board for Soil Research and Management: Bangkok, Thailand, 1991; Volume 2, pp. 175–203.

9. Tugel, A.J.; Herrick, J.E.; Brown, J.R.; Mausbach, M.J.; Puckett, W.; Hipple, K. Soil change, soil survey, and natural resources decision making: A blueprint for action. *Soil Sci. Soc. Am. J.* **2005**, *69*, 738–747.

10. Food and Agricultural Organization of the United Nations (FAO) and Intergovermental Panel on Soils (ITPS). *Status of the World's Soil Resources—Main Report*; FAO/ITPS: Rome, Italy, 2015; pp. 1–648.

11. Food and Agricultural Organization of the United Nations (FAO) and Intergovermental Panel on Soils (ITPS). 8 governance and policy responses to soil change. In *Status of the World's Soil Resources—Main Report*; FAO/ITPS: Rome, Italy, 2015; pp. 223–241.

12. Food and Agricultural Organization of the United Nations (FAO) and Intergovermental Panel on Soils (ITPS). 14 regional assessment of soil changes in North America. In *Status of the World's Soil Resources—Main Report*; Pennock, D., Ed.; FAO/ITPS: Rome, Italy, 2015; pp. 442–475.

13. Food and Agricultural Organization of the United Nations (FAO) and Intergovermental Panel on Soils (ITPS). 15 regional assessment of soil change in the southwest pacific. In *Status of the World's Soil Resources—Main Report*; McKenzie, N.J., Ed.; FAO/ITPS: Rome, Italy, 2015; pp. 476–516.

14. Folke, C.; Carpenter, S.R.; Walker, B.; Scheffer, M.; Chapin, T.; Rockström, J. Resilience Thinking: Integrating Resilience, Adaptability and Transformability. Available online: http://www. ecologyandsociety. org/vol15/ iss4/art20/(accessed on 24 March 2016).

15. Walker, B.H.; Carpenter, S.R.; Rockstrom, J.; Crépin, A.S.; Peterson, G.D. Drivers, "slow" variables, "fast" variables, shocks, and resilience. *Ecol. Soc.* **2012**.

16. Williams, J. Soils governance in Australia: Challenges of cooperative federalism. *Int. J. Rural Law Policy* **2015**, *1*, 1–12.

17. Baumhardt, R.L.; Stewart, B.A.; Sainju, U.M. North American soil degradation: Processes, practices and mitigating strategies. *Sustainability* **2015**, *7*, 2936–2960.

18. Cary, J.; Roberts, A. The limitations of environmental management systems in Australian agriculture. *J. Environ. Manag.* **2011**, *92*, 878–885.

19. Benton, J. Soil analysis (testing). In *Laboratory Guide for Conducting Soil Tests and Plant Analysis*; CRC Press: Boca Raton, FL, USA, 2001; pp. 11–190.

20. Rayment, G.E.; Higginson, F.R. *Australian Laboratory Handbook of Soil and Water Chemical Methods*; Inkata Press: Melbourne, Australia, 1992.

21. Rayment, G. Soil and plant testing in Australia—The players, clients, methods and issues. In *Tools for Nutrient and Pollutant Management—Applications to Ariculture and Environmental Quality*; Currie, L.D., Hanly, J.A., Eds.; Fertlizer and Lime Research Centre: Massey University, New Zealand, 2004; pp. 9–23.

22. Westerman, R.L. *Soil Testing and Plant Analysis*; Soil Science Society of America: Madison, WI, USA, 1990.

23. Peverill, K.I. Soil testing and plant analysis in Australia. *Aust. J. Exp. Agric.* **1993**, *33*, 963–971.

24. Peverill, K.I.; Sparrow, L.A.; Reuter, D.J. *Soil Analysis: An Interpretation Manual*; CSIRO Publishing: Melbourne, Australia, 1999.

25. Department of Agriculture and Water Resources. Monitor. Available online: http://data.daff.gov.au/monitor/explore.html (accessed on 24 March 2016).

26. Australian Bureau of Statistics (ABS). 4627.0-Land Management and Farming in Australia. Available online: http://www.abs.gov.au/ausstats/abs@.nsf/mf/4627.0 (accessed on 24 March 2016).

27. United States Department of Agriculture, Economic Research Service. ARMS Farm Financial and Crop Production Practices. Available online: http://www.ers.usda.gov/data-products/arms-farm-financial-and-crop-production-practices/tailored-reports-farm-structure-and-finance.aspx (accessed on 24 March 2016).

28. Andrews, S.S.; Flora, C.B.; Mitchell, J.P.; Karlen, D.L. Growers' perceptions and acceptance of soil quality indices.*Geoderma* **2003**, *114*, 187–213.

29. National Resource Conservation Service National Soil Survey Center. Soil quality for environmental health. Available online: http://soilquality.org/home.html (accessed on 24 March 2016).

30. Australian Bureau of Statistics (ABS). *4627 Land Management and Farming in Australia 2007–2008*; ABS: Canberra, Australia, 2009.

31. Australian Bureau of Statistics (ABS). *4627 Land Management and Farming in Australia, 2009–2010*; ABS: Canberra, Australia, 2011.

32. Australian Bureau of Statistics (ABS). *4627 Land Management and Farming in Australia 2011–2012*; ABS: Canberra, Australia, 2013.

33. Australian Bureau of Statistics (ABS). *4627 Land Management and Farming in Australia 2012–2013*; ABS: Canberra, Australia, 2014.

34. Bennett, J.M.; Lobry de Bruyn, L.A.; University of Southern Queensland, Towoomba, Australia. Personal Communication, 2015.

35. Bennett, J.M. *Combating sodicity in the Lachlan and Macquarie valleys of New South Wales*; Sydney University: Sydney, Australia, 2011.

36. Kearns, S.; Umbers, A. *GRDC Farm Practices Baseline Report 2010*; GRDC: Canberra, Australia, 2010.

37. Edwards, J.; Umbers, A.; Wentworth, S. *GRDC Farm Practices Survey Report 2012*; GRDC, Kingston, ACT: Barton, Australia, 2012.

38. Australian Bureau of Statistics (ABS). *4627 Land Management and Farming in Australia 2013–2014*; ABS: Canberra, Australia, 2015.

39. Lambert, D.; Sullivan, P.; Claassen, R.; Foreman, L. *Conservation-Compatible Practices and Programs: Who Participates?*Economic Research Service/United States Department of Agriculture: Washington, DC, USA, 2006; pp. 1–42.

40. Foreman, L. Characteristics and production costs of U.S. Corn farms. *Statistical Bulletin 974*; United States Department of Agriculture, Economic Research Service: Washington, DC, USA, 2001; pp. 1–28.

41. Foreman, L. Characteristics and production costs of U.S. Corn farms, 2001. *Economic Information Bulletin, No 7*; United States Department

of Agriculture, Economic Research Service: Washington, DC, USA, 2006; pp. 1–47.

42. Foreman, L. Characteristics and production costs of U.S. Corn farms, including organics, 2010. *Economic Information Bulletin 128*; United States Department of Agriculture, Economic Research Service: Washington, DC, USA, 2014; pp. 1–37.

43. Huang, W.; Beckman, J. Agricultural production management. *Agricultural Resources and Environmental Indicators*, 2012 ed.; Osteen, C., Gottlieb, J., Vasavada, U., Eds.; USDA, ERS: Washington, DC, USA, 2012; pp. 25–28.

44. Plank, O.C. *Soil, Plant and Animal Waste Water Analysis: Status Report for the United States*; Soil and Plant Analytical Council Incorporated: Lincoln, NE, USA, 1998.

45. United States Department of Agriculture, NASS. *2012 Census of Agriculture: Preliminary Report Highlights: U.S. Farms and Farmers*; United States Department of Agriculture, National Agricultural Statistics Service: Washington, DC, USA, 2014; pp. 1–4.

46. Fixen, P.E. Soil test levels in North America. *Better Crop.* **2002**, *86*, 12–15.

47. United States Department of Agriculture (USDA). *Summary Report: 2010 National Resources Inventory*; 2013.

48. Miller, R.; Lobry de Bruyn, L.A.; Colorado State University, Fort Collins, CO, USA. Personal Communication, 2016.

49. Australian Bureau of Statistics (ABS). *7121.0 Agricultural Commodities, Australia, 1997–1998*; ABS: Canberra, Australia, 1999.

50. Australian Bureau of Statistics (ABS). *7121.0 Agricultural Commodities, Australia, 2002–2003*; ABS: Canberra, Australia, 2004.

51. Commonwealth of Australia. *The National Soil Research, Development and Extension Strategy, Securing Australia's Soil, for Profitable Industries and Healthy Landscapes*; Department of Agriculture Forestry and Fisheries DAFF: Canberra, Australia, 2014; p. 75.

52. Reuter, D.J.; Hannam, R.J. *Soil and Plant Testing in Australia*; Australian Society of Agronomy: Melbourne, Australia, 1987; pp. 34–46.

53. Wilson, B.; Lobry de Bruyn, L.A.; University of New England, Armidale, Australia. Personal Communication, 2015.

54. Oliver, M.; Ashton, D.; Hodges, A.; Mackinnon, D. *Farmers' Use of Sustainable Management Practices*; ABARE Report for the National Land and Water Resources Audit: Canberra, Australia, 2009.

55. Mues, C.; Chapman, L.; Van Hilst, R. *Promoting Improved Land Management Practices on Australian Farms: A Survey of Landcare and Land Management Related Programs*; ABARE: Canberra, Australia, 1998.

56. Mues, C.; Roper, H.; Ockerby, J. *Survey of Landcare and Land Management Practices: 1992–1993*; ABARE: Canberra, Australia, 1994.

57. Hodges, A.; Goesch, T. *Australian Farms: Natural Resource Management in 2004–2005*; ABARE: Canberra, Australia, 2006.

58. Campbell, A. *Managing Australia's Soils: A Policy Discussion Paper. National Committee on Soils and Terrain*; DAFF: Canberra, Australia, 2008.

59. Ecker, S.; Thompson, L.; Kancans, R.; Stenekes, N.; Mallawaaraachchi, T. *Drivers of Practice Change in Land Management in Australian Agriculture: Results of a National Farm Survey*; ABARES: Canberra, Australia, 2012; p. 63.

60. Ribaudo, M. Conservation policies: Working-lands conservation programs. *Agricultural Resources and Environmental Indicators*, 2012 ed.; Osteen, C., Gottlieb, J., Vasavada, U., Eds.; United States Department of Agriculture, Economic Research Service: Washington, DC, USA, 2012; pp. 47–48.

61. National Wildlife Federation (NWF). *A Roadmap to Increased Cover Crop Adoption*; NWF: Merrifield, VA, USA, 2012; pp. 1–12.

62. Romig, D.E.; Garlynd, M.J.; Harris, R.F.; McSweeney, K. How farmers assess soil health and quality. *J. Soil Water Conserv.* **1995**, *50*, 229–236.

63. Liebig, M.A.; Doran, J.W. Evaluation of farmers' perceptions of soil quality indicators. *Am. J. Altern. Agric.* **1999**, *14*, 11–21.

64. Lobry de Bruyn, L.A.; Abbey, J.A. Characterisation of farmers' soil sense and the implications for on-farm monitoring of soil health. *Aust. J. Exp. Agric.* **2003**, *43*, 285–305.

65. Speirs, S.D.; Reuter, D.J.; Peverill, K.I.; Brennan, R.F. Making better fertiliser decisions for cropping systems in Australia: An overview. *Crop Pasture Sci.* **2013**, *64*, 417–423.

66. Ingram, J.; Fry, P.; Mathieu, A. Revealing different understandings of soil held by scientists and farmers in the context of soil protection and management. *Land Use Policy* **2010**, *27*, 51–60.

67. Ingram, J. Are farmers in England equipped to meet the knowledge challenge of sustainable soil management? An analysis of farmer and advisor views. *J. Environ. Manag.* **2008**, *86*, 214–228.

68. Kelly, B.; Allan, C.; Wilson, B.P. Soil indicators and their use by farmers in the Billabong catchment, southern New South Wales. *Soil Res.* **2009**, *47*, 234–242.

69. Commonwealth of Australia. *Managing Natural Resources in Rural Australia for a Sustainable Future: A Discussion Paper for Developing a National Policy*; Department of Agriculture, Fisheries and Forestry, Australia: Canberra, Australia, 1999.

70. Bennett, J.M.; Cattle, S.R. Adoption of soil health improvement strategies by Australian farmers: I. Attitudes, management and extension implications. *J. Agric. Educ. Ext.* **2013**, *19*, 407–426.

71. Bennett, J.M.; Cattle, S.R. Adoption of soil health improvement strategies by Australian farmers: II. Impediments and incentives. *J. Agric. Educ. Ext.* **2014**, *20*, 107–131.

72. Australian Bureau of Statistics (ABS). *4627 Land Management and Farming in Australia, 2010–2011*; ABS: Canberra, Australia, 2012.

73. Fixen, P.E. Soil test levels in North America. *Better Crop.* **1998**, *82*, 16–18.

74. Fixen, P.E.; Bruulsema, A.M.; Johnston, T.W.; Mikkelsen, R.L.; Murrell, T.S.; Snyder, C.S.; Stewart, W.M. *The Fertility of North American Soils*; Potash & Phosphate Institute: Norcross, GE, USA, 2006.

75. Fixen, P.E.; Bruulsema, T.W.; Jensen, T.L.; Mikkelsen, R.; Murrell, T.S.; Phillips, S.B.; Rund, Q.; Stewart, W.M. The fertility of North American soils. *Better Crop.* **2010**, *94*, 6–8.

76. Potash and Phosphate Institute (PPI). Soil test summaries: Phosphorus, Potassium and pH. *Better Crop.* **1990**, *74*, 16–18.

77. Potash and Phosphate Institute (PPI). Soil test summaries: Phosphorus, Potassium, and pH. *Better Crop.* **1994**, *78*, 14–17.

78. International Plant Nutrition Institute (IPNI). *Soil Test Levels in North America*; IPNI: Peachtree Corners, GA, USA, 2010.

79. Fixen, P.E. Soil test levels in North America. *Better Crop.* **2006**, *90*, 4–7.

80. Ward Laboratories Incorporated. Ward laboratories, Inc. Available online: https://producers.wardlab.com/default.aspx(accessed on 24 March 2016).

81. United States Department of Agriculture Natural Resources Conservation Service. CSP Enhancement Activity Job Sheets. Available online: http://www.nrcs.usda.gov/wps/portal/nrcs/detailfull/national/programs/financial/csp/?cid=stelprdb1265825 (accessed on 24 March 2016).

82. Cornell University College of Agriculture and Life Sciences. Comprehensive Assessment of Soil Health-The Cornell Framework Manual. Available online: http://soilhealth.cals.cornell.edu/ (accessed on 21 March 2016).

83. USDA NRCS. *Soil Quality Enhancement Activity—SQL15—Utilize the Soil Health Nutrient Tool to Assess Soil Nutrient Pools*; USDA: Washington, DC, USA, 2014.

84. Porsborg, D.; Lobry de Bruyn, L.A.; Farmer, Oliver County, ND, USA. Personal Communication, 2015.

85. Martin, P.; Riley, D.; Ashton, D.; Delforce, J.; Loggia, D. *Australian Grains Industry 1998: Report of the Australian Agricultural and Grazing Industries Survey of Grain Producing Farms*; ABARE: Canberra, Australia, 1998.

86. Bosch, O.J.H.; Allen, W.J.; Gibson, R.S. Monitoring as an integral part of management and policy making. In Proceedings of the Symposium on Resource Management: Issues, Visions, Practice, Lincoln, New Zealand, 5–8 July 1996; pp. 12–21.

87. Brown, R.; USDA ERS, Washington, DC, USA. Unpublished work. 2015.

88. Barson, M.; Mewett, J.; Paplinksa, J. *Land Management Practice Trends in Australia's Broadacre Cropping Industries*; DAFF: Canberra, Australia, 2012.

89. Pannell, D.J.; Marshall, G.R.; Barr, N.; Curtis, A.; Vanclay, F.; Wilkinson, R. Understanding and promoting adoption of conservation practices by rural landholders. *Aust. J. Exp. Agric.* **2006**, *46*, 1407–1424.

90. Kancans, R.; Ecker, S.; Duncan, A.; Stenekes, N.; Zobel-Zubrzycka, H. *Drivers of Practice Change in Land Management in Australian Agriculture, Synthesis Report—Stages I, II and III*; ABARES: Canberra, Australia, 2014.

91. Llewellyn, R.S.; D'Emden, F. *Adoption of No-Till Cropping Practices in Australian Grain Growing Regions*; Australian Government, Grains Research and Development Corporation: Kingston, Australia, 2010.

92. D'Emden, F.H.; Llewellyn, R.S.; Burton, M.P. Factors influencing adoption of conservation tillage in Australian cropping regions. *Aust. J. Agric. Resour. Econ.* **2008**, *52*, 169–182.

93. Mitchell, J.; Andrews, S.S.; Kearney Agricultural Research and Extension Center, Parlier, CA, USA. Personal Communication, 2010.

94. Department of Primary Industries Victoria. *Soil Heallth Management Plan: Concept, Process and Pilot Study with Mid-Loddan Sub-Catchment Group*; Department of Primary Industries: Melbourne, Australia, 2009.

95. Tiemann, L.K.; Grandy, A.S.; Atkinson, E.E.; Marin-Spiotta, E.; McDaniel, M.D. Crop rotational diversity enhances belowground communities and functions in an agroecosystem. *Ecol. Lett.* **2015**, *18*, 761–771.

96. Wallander, S. While crop rotations are common, cover crops remain rare. *Amber Waves*; USDA ERS: Washington, DC, USA, 2013.

97. Conservation Effects Assessment Project (CEAP). *Assessment of the Effects of Conservation Practices on Cultivated Cropland in the Upper Mississippi River Basin*; United States Department of Agriculture, Natural Resources Conservation Service: Washington, DC, USA, 2012.

98. Singer, J.W. *Cover Crops in the Corn Belt: Survey Finds Underused Potential as Conservation Tool*; Leopold Center for Sustainable Agriculture: Ames, IA, USA, 2006; p. 3.

99. Conservation Technology Innovation Centre/Sustainable Agriculture Research and Extension CTIC/SARE. *2012–2013 Annual Report—Cover Crop Survey*; CTIC/SARE: West Lafayette, IN, USA, 2014; pp. 1–50.

100. Conservation Technology Innovation Centre/Sustainable Agriculture Research and Extension CTIC/SARE. *2014–2015 Annual Report—Cover Crop Survey*; CTIC/SARE: West Lafayette, IN, USA, 2015; pp. 1–45.

101. Australian Bureau of Statistics (ABS). *Australian Social Trends December 2012: Australian Farming and Farmers*; Commonwealth of Australia: Canberra, Australia, 2012.

102. Cary, J.; Webb, T.; Barr, N. *Understanding Landholders' Capacity to Change to Sustainable Practices*; Bureau of Rural Sciences: Canberra, Australia, 2002.

103. Gazey, C.; Oliver, Y.; Fisher, J.; Andrew, J.; Carr, S. *20 Years of soil Acidity RD and E in Western Australia: What Have We Learnt?* Western Australia Department of Agriculture: Perth, Australia, 2014; pp. 1–5.

104. Australian Bureau of Statistics (ABS). *4620 Natural Resource Management on Australian Farms*; ABS: Canberra, Australia, 2008.

105. Australian Bureau of Statistics (ABS). *4620 Natural Resource Management on Australian Farms 2004–2005*; ABS: Canberra, Australia, 2007.

106. Bramley, R.; Cook, S. Some comments on the utility of soil test data for fertility management and resource assessment. In *ACLEP Newsletter*; CSIRO: Canberra, ACT, 2000; pp. 27–34.

107. Nickerson, C.M.; Morehart, T.; Kuethe, J.; Beckman, J.; Ifft, J.; Williams, R. Trends in U.S. Farmland values and ownership. *Economic Information Bulletin Number 92*; USDA Economics Research Service: Washington, DC, USA, 2012.

108. Soil Health Partnership. Tim Smith Eagle Grove, Iowa. Available online: http://www.soilhealthpartnership.org/meet-tim-smith.html (accessed on 24 March 2016).

109. Rogers, E.M. *Diffusion of Innovations*; Free Press: New York, NY, USA, 2003.

110. Johnson, G.V.; Westerman, R.L. Improving soil testing credibility. *Commun. Soil Sci. Plant Anal.* **1980**, *11*, 251–255.

111. Loescher, H.; Ayres, E.; Duffy, P.; Luo, H.; Brunke, M. Spatial variation in soil properties among North American ecosystems and guidelines for sampling designs. *PLoS ONE* **2014**.

112. Cook, S.; Bramley, R. Coping with variability in agricultural production—Implications for soil testing and fertiliser management. *Commun. Soil Sci. Plant Anal.* **2000**, *31*, 1531–1551.

113. Lehman, R.; Cambardella, C.; Stott, D.; Acosta-Martinez, V.; Manter, D.; Buyer, J.; Maul, J.; Smith, J.; Collins, H.; Halvorson, J.; *et al.* Understanding and enhancing soil biological health: The solution for reversing soil degradation.*Sustainability* **2015**, *7*, 988–1027.

114. Wall, D.H.; Nielsen, U.N.; Six, J. Soil biodiversity and human health. *Nature* **2015**, *528*, 69–76.

115. USDA ARS NRCS SQI. *Soil Quality Test Kit*; USDA: Washington, DC, USA, 2001; p. 82.

116. USDA NRCS SQI. *Soil Quality Card Design Guide: A Guide to Develop Locally Adapted Conservation Tools*; USDA: Washington, DC, USA, 1999; p. 115.

117. United States Department of Agriculture, National Resources Conservation Service. Soil Health Card. Available online: http://www.nrcs.usda.gov/wps/portal/nrcs/detail/soils/health/assessment/?cid=nrcs142p2_053871 (accessed on 24 March 2016).

118. Liebig, M.A.; Doran, J.W.; Gardner, J.C. Evaluation of a field test kit for measuring selected soil quality indicators.*Agron. J.* **1996**, *88*, 683–686.

119. Karlen, D.L.; Wienhold, B.J.; Kang, S.; Zobeck, T.M.; Andrews, S.S. Chapter 3: Indices for soil management decisions.*Soil Management: Building a Stable Base for Agriculture*; Hatfield, J., Sauer, T.J., Eds.; American Society of Agronomy and Soil Science Society of America: Madison, WI, USA, 2011.

120. McKenzie, N.; Lobry de Bruyn, L.A.; CSIRO Land and Water, Canberra, Australia. Personal Communication, 2015.

121. United States Department of Agriculture Natural Resources Conservation Service. Web Soil Survey. Available online:http://websoilsurvey.sc.egov.usda.gov/App/HomePage.htm (accessed on 24 March 2016).

122. USDA NASS. *Farm Computer Usage and Ownership*; USDA NASS: Washington, DC, USA, 2013; pp. 1–29.

123. eXtensionAUS. Australia's Research-Based Learning Network. Available online: http://www.extensionaus.com.au/(accessed on 24 March 2016).

124. eXtension. For extension Professionals and the Public They Serve. Available online: https://extension.org/ (accessed on 24 March 2016).

125. Beange, L.; Lobry de Bruyn, L.A.; Department of Primary Industries, NSW, Dubbo, Australia. Personal Communication, 2015.

126. Land & Water Australia. Knowledge for Managing Australian Landscapes 1990–2009. Available online: http://lwa.gov.au/ (accessed on 24 March 2016).

127. Land & Water Australia. National Land and Water Resources Audit. Available online: http://lwa.gov.au/programs/national-land-and-water-resources-audit (accessed on 24 March 2016).

128. Food and Agricultural Organization of the United Nations (FAO). FAOKnowledge. Available online: https://twitter.com/FAOKnowledge (accessed on 24 March 2016).

129. United States Department of Agriculture Natural Resources Conservation Service North Dakota. Unlock the Secrets in the Soil. Available online: http://www.nrcs.usda.gov/wps/portal/nrcs/main/nd/soils/health/ (accessed on 24 March 2016).

130. United States Department of Agriculture Natural Resources Conservation Service. Minnesota Farm Uses Conservation to Make Each Acre Count. Available online: http:// www.nrcs.usda. gov/wps/portal /nrcs/detail/national/newsroom/features/?cid=stelprdb1249011 (accessed on 24 March 2016).

131. UCDavis California Soil Resource Lab. Soil Web Apps. Available online: http://casoilresource.lawr.ucdavis.edu/soilweb-apps/ (accessed on 24 March 2016).

132. Soil Quality Pty Ltd. Welcome to the Soil Quality Website. Available online: http://www.soilquality.org.au/ (accessed on 24 March 2016).

133. Hoover, D.; Lobry de Bruyn, L.A.; USDA National Soil Survey Centre, Lincoln, NE, USA. Personal Communication, 2015.

134. Hunt, W.; Birch, C.; Vanclay, F.; Coutts, J. Recommendations arising from an analysis of changes to the Australian agricultural research, development and extension system. *Food Policy* **2014**, *44*, 129–141.

135. Thuo, M.; Bell, A.; Bravo-Ureta, B.; Lachaud, M.; Okello, D.; Okoko, E.; Kidula, N.; Deom, C.; Puppala, N. Effects of social network factors on information acquisition and adoption of improved groundnut varieties: The case of uganda and kenya. *Agric. Hum. Values* **2014**, *31*, 339–353.

136. Soils for Life Pty Ltd. Soils for Life. Available online: http://www.soilsforlife.org.au/about.html (accessed on 24 March 2016).

137. SoilCare Inc. Available online: http://www.soilcare.org/about.html (accessed on 24 March 2016).

138. Soil Health Partnership. Available online: http://www.soilhealthpartnership.org/ (accessed on 24 March 2016).

139. USDA NASS. *Conservation Effects Assessment Project (CEAP) 2004 Survey*; USDA NASS: Washington, DC, USA, 2004.

140. Arbuckle, J.G.; Ferrell, J. *Attitudes toward Cover Crops in Iowa: Benefits and Barriers*; Iowa State University Extension and Outreach: Ames, IA, USA, 2012.

141. Nelson, R.; Alexander, F.; Elliston, L.; Blias, A. *Natural Resource Management on Australian Farms*; ABARE: Canberra, Australia, 2004.

Chapter 9

Use of different organic fertilizers on soil fertility improvement, growth and head yield parameters of cabbage (*Brassica oleraceae* L)

Emmanuel Ibukunoluwa Moyin-Jesu

Agronomy Department, Federal College of Agriculture, Akure, Ondo, Nigeria

ABSTRACT

Purpose Field experiment was carried out in Akure in the rainforest zone of Nigeria to determine the effect of poultry manure, wood ash and rice bran on the soil fertility improvement, growth and head yield of cabbage (*Brassica oleraceae* L) in 2011 and 2012 cropping seasons.

Methods The three organic fertilizer treatments were applied each at 6 t/ha with a reference treatment NPK 15-15-15 applied at 300 kg/ha and a control treatment arranged in a randomized complete block design with three replicates.

Results The results showed significant increases ($P < 0.05$) in the growth and head yield parameters of cabbage crop under the different organic fertilizers compared to the control treatment. The application of poultry manure resulted in the highest values of cabbage plant height, stem girth, leaf number, leaf area, head weight, head length and head girth followed by NPK 15-15-15, wood ash and rice bran, respectively. Cabbage head weight, head girth, head length, plant height, stem girth, leaf number and leaf area increased by 17, 18, 8,17, 19, 10 and 16 %, respectively, with application of poultry manure compared to NPK fertilizer. Also, these parameters increased by 23, 21, 13 29, 25, 17 and 45 % compared to the wood ash treatment. The highest values of soil pH, Ca and Mg were obtained under wood ash application, while the highest values of soil O.M and moderate values of soil P, K, Ca and Mg were obtained under poultry manure application. Moderate values of soil pH, K, Ca and low % N were also recorded under rice bran treatment. However, the highest K/Ca, K/Mg and P/Mg ratios of 93.1, 74.1 and 572.1, respectively, were obtained under NPK 15-15-15 fertilizer application compared to 2:1 K/Ca, 2:1 K/Mg and 16:1 P/Mg in rice bran treatment.

Conclusion The poultry manure applied at 6 t/ha gave the best results in improving soil fertility, growth and head yield of cabbage, and this was because of its balanced nutrient composition and the least C/N ratio.

KEYWORDS

Cabbage (*Brassica oleraceae*) Head yield and growth parameters Organic fertilizers and soil fertility improvement

1. INTRODUCTION

Cabbage is a leafy vegetable belonging to the Brassicaceae family which is originated from California. Hague (2006) cabbage is also an herbaceous biennial plant that produces “heads”, and it has short stem upon which is crowned with a mass of green leaves (John 1997).

Cabbage is consumed raw or cooked with other vegetables; it contains 93 ml water, 15 g protein, 0.2 g fat, 4.0 g carbohydrate, 4 g calcium and 0.5 g iron/100 g sample (Moamogwe 1995). In addition, it has high medicinal value because of its rich content in vitamins C, E and K as well as its antioxidant components which suppressed cancerous cell growth (Tindall

2000). It is one important vegetable in healthy diets of people around the world.

In spite of the economic, nutritional and health importance of cabbage to the national development, its optimum production has not been achieved due to its difficult agronomic characteristics, decline in soil fertility, lack of techniques of growing the crop in both the nursery and field. However, efforts made to increase the soil fertility through the use of inorganic fertilizers are limited due to high cost of purchase and continuous use which destroyed soil properties (Moyin-Jesu 2008).

Besides, the climatic condition for growing cabbage in temperate region is between 20 and 25 °C and adaptation of the crop to Nigeria, especially south-west, is difficult because temperature is between 29 and 37 °C. In other parts of the country (Northern region), the temperature ranged between 36 and 41 °C except the Jos Plateau where temperature is about 24–27 °C.

Therefore, there is justification to adapt the cultivation of cabbage to the climatic condition in Nigeria, especially in south-west, as well as adopting the use of low-cost organic fertilizers for the crop cultivation. Kwari et al. (1991) found that the incorporation of grass cow dung into the soils in Borno state improved the cation exchange capacity of the soils. Solomon and Ogeh (1995) indicated that leguminous materials and rice husk supplied mainly N, P, K, Zn, Fe, Cu, Mn and B to the soil which NPK 15-15-15 fertilizer did not possess.

Ojeniyi et al. (2001) also reported that application of wood ash to the vegetable crops improved the yield of crops significantly and increased the soil nutrients such as N, P, K, Ca, Mg in a trial conducted in south-west Nigeria. Awodun (2007) also reported that sawdust applied to soil significantly increased soil and leaf N, P, K Ca and Mg contents and number of pods, pod weight, number of branches, number of leaves and grain yield of cowpea.

Moyin-Jesu (2013a) reported that the modified neem extracts (neem leaf + wood ash extracts) increased significantly the garden egg fruit weight, fruit length and fruit diameter by 42, 24 and 13 % compared to NPK 15-15-15 fertilizer. He also observed that continuous application of NPK 15-15-15 fertilizer decreased soil pH, Ca, Mg and O.M as well as having highest K/Ca, K/Mg and P/Mg ratio caused soil nutrient imbalance.

Moyin-Jesu and Ogochukwu (2014) also reported that poultry manure applied at 6 t/ha increased the soil N, P, K, Ca and Mg concentrations and growth parameters of coconut seedlings. He also reported that the organic materials applied (wood ash, rice bran and so firth) have beneficial residual effects on soil properties which are in line with growing concern of using environment friendly fertilizer.

Furthermore, Obi and Ofonduro (1997) and Moyin-Jesu (2007) also reported that problems associated with continuous use of chemical fertilizers included nutrient imbalance, increased soil acidity, degradation in soil physical properties and loss of organic matter. Hence, the tendency to supply all plant nutrients through chemical fertilizer should be reconsidered in the future because of the deleterious effect on soil productivity on a long-term basis.

Finally, Afshar et al. (2014) reported that soil organic amendments improved the seed yield and flavonolignan production of milk thistle (*Silyburn marianum* L, Gaertin) compared to chemical fertilizers. Critical review of literature revealed that there is scarcity of research report on the use of wood ash, rice bran and poultry manure in growing cabbage except the works of Moyin-Jesu and Odewande (2013b); Ijoyah and Sophie (2009) and Mohammed and Solaiman (2012) working on varietal trial of cabbage and tillage influence on cabbage yield.

The choice of wood ash, rice bran and poultry manure as source of organic materials in the research study was because of their availability and abundance in large quantities from the experimental location and its surrounding communities where majority of the farmers are predominantly cassava, rice growers with associated processing mills and poultry producers. Moyin-Jesu and Ojeniyi (2006) also reported that application of rice bran wood ash and spent grain significantly reduced soil bulk density and increased porosity. Hence, the use of wood ash and rice bran as organic amendments would also have strong beneficial improvement on soil physical properties and base saturation.

1.1 Purpose of the experiment

The following research questions would be answered for the experiment: (1) Is there any significant difference between applied organic fertilizers on growth and head yield parameters of cabbage? (2) Is there any significant difference between the applied organic fertilizers and post-cropping soil properties?

The objectives of the research were to determine the effect of wood ash, rice bran and poultry manure on the growth and yield parameters of cabbage and on post-cropping soil properties.

2. MATERIALS AND METHODS

2.1 Description of the study area

Field and nursery experiments were carried out at Akure in the rainforest zone of Nigeria in 2011 and were repeated in 2012 to validate the results. The soil of the experimental site is loamy sand, skeletal, kaolinitic, isohyperthermic Oxic paleustalf (Alfisol) Soil Survey Staff (1999). The annual rainfall of the area is between 1000 and 2060 mm and the annual temperature is between 29–37 °C as presented in Table 1.

2.2 Pre-cropping soil sampling and analysis

Thirty core samples were collected from 0 to 15 cm depth, bulked together, air-dried, sieved with 2 mm and ready for routine analysis.

The soil pH (1:1 soil/water and 1:2 soil/0.01 M $Cacl_2$) was read on pH meter (Crockford and Nowel 1956), while organic matter was determined using wet oxidation method through chromic acid digestion (Walkley and Black 1934). Soil P was extracted by Bray P1 extractant, and the extract was developed on Murphy blue colouration and determined on a spectronic 20 (Murphy and Riley 1962).

Soil K, Ca, Mg and Na were extracted with 1 M NH_40AC pH7, and their contents K, Ca and Na were read on the flame photometer (Jackson 1958), while Mg content was determined using the atomic absorption spectrophotometer.

Table 1. Monthly rainfail and temprature data for Akure in 2011 and 2012

Months	Rainfall (mm)		Temperature (0 °C)		Modified temp. (0 °C) of the environment under shade structure during the experiment (March–July)	
	2011	2012	2011	2012	2011	2012
January	1.90	–	33.70	36.30	–	–
February	41.10	34.80	36.10	35.60	–	–
March	127.10	40.00	35.00	37.10	22.10	22.60
April	230.00	81.80	29.10	32.20	23.10	23.40
May	222.30	156.60	30.30	29.30	22.80	22.60
June	162.00	195.80	32.20	31.20	22.50	22.30
July	169.00	423.00	29.10	29.20	22.30	22.00
August	169.00	682.90	31.40	30.10	22.90	22.70
September	378.00	240.90	30.70	32.30	–	–
October	141.90	111.60	33.10	31.10	–	–
November	51.30	93.20	33.60	35.30	–	–
December	–	–	32.70	34.60	–	–

Source: meteorological station Federal College of Agriculture, Akure

The soil nitrogen was determined using the micro-Kjeldahl method (Jackson 1964), while the micronutrients (Fe, Cu, Zn and Mn) were extracted with 0.1 M, HCl and read on atomic absorption spectrophotometer. Particle size analysis was carried out using hydrometer method (Bouycous 1951).

Table 2 shows the soil chemical composition before planting. The soil P 5.20 mg/kg is lower than 10 mg/kg P recommended for crop production in south-west Nigeria (Agboola and Corey 1973). The organic matter was also below 3 % critical level for crop production (Agboola and Corey 1973).

Table 2. Soil properties before planting

Soil properties	Values
Soil pH (H_2O)	5.82
Soil pH (0.01 M $CaCl_2$)	5.34
Organic matter (%)	0.60
Nitrogen (%)	0.08
Available P (mg/kg)	5.20
Exchangeable bases	
K^+ (mmol/kg)	0.11
Ca^{2+} (mmol/kg)	0.13
Mg^{2+} (mmol/kg)	0.08
Na^{2+} (mmol/kg)	0.11
Exchangeable acidity	
H^+ (mmol/kg)	4.10
Al^{3+} (mmol/kg)	1.30
Micronutrients	
Fe (mg/kg)	8.30
Zn (mg/kg)	3.83
Cu (mg/kg)	2.20
Mn (mg/kg)	1.85
Particle size analysis	
Sand (%)	79.30
Silt (%)	14.70
Clay (%)	6.00

Soil N content was 0.08 % which is lower than 0.15 % N critical level for crops recommended by Sobulo and Osiname (1981). The exchangeable bases (K, Ca, Mg and Na) were below 0.20 mmol/kg recommended by Folorunso et al. (2000). The soil pH 5.82 showed that the soil is slightly acidic. The soil textural class is loamy sand.

2.3 Sources and processing of organic fertilizers used for the experiments

Rice bran and wood ash were obtained from the large-scale processing unit of Federal College of Agriculture, Akure, which processed cassava tubers from 10,000 ha of cassava farm and paddy rice from 5000 hectares of rice plantation, while the poultry manure was also obtained from the 10,000 poultry birds in the livestock unit of the same institution. NPK 15-15-15 was purchased from Agricultural Development Programme, while the imported cabbage seeds (Benelli cultivar) were purchased from a certified seed company Agro-Farm Enterprises, Akure.

The organic materials were processed, wood ash was sieved with 2 mm sieve to remove pebbles, while rice bran was partially composted for four weeks to reduce C/N ratio. Poultry manure was cured before application.

2.4 Chemical analysis of the organic materials used

Two grammes each of the processed forms of the organic materials was analysed. The N content was determined by the Kjeldahl method (Jackson 1964), while the determination of other nutrients such as P, K, Ca, Mg was done using the wet digestion method based on 25-5-5 ml of HNO_3-H_2SO_4-$HClO_4$ acids (AOAC 1970).

Table 3 shows the chemical analysis of the organic fertilizer used for the experiment. Wood ash had the highest values of K, Ca and Mg followed by poultry manure and rice bran, respectively. Rice bran had the least values of N, P, K, Ca and Mg and highest value of C/N ratio. Poultry manure also had the highest value of N, P and least C/N ratio compared to others.

Table 3. Chemical analysis of the organic fertilizer used for the experiment

Treatments	N (%)	C (%)	C/N	P (mg/kg)	°K (%)	Ca (%)	Mg (%)
Poultry manure	4.32	30.00	6.93	385.00	9.70	3.30	4.10
Wood ash	1.53	18.00	11.76	86.00	23.02	9.40	8.52
Rice bran	0.60	14.00	23.33	56.00	7.93	0.12	1.80

2.5 Nursery establishment for cabbage seedlings

Land clearing and packing of debris were carried out followed by shade construction made of erected bamboo poles and top covered with moderate number of palm fronds. Twelve nursery beds with a size of 1 × 2 m^2 each were prepared. The cabbage seeds were sown on March manually into rows spaced at 30 cm apart, irrigated twice a day (morning and evening) to ensure good germination and establishment. Germination of the seeds occurred 8 days after planting and seedlings were nursed for 2 weeks before transplanting. The nursery establishment is important because directly sown seeds will not germinate; hence, the erection of shade structure in the experiment was to control the prevailing temperature of the environment for seeds germination.

2.6 Field experiment

The experimental site was cleared, ploughed, harrowed and divided into different plots. Each plot size was 4 × 4 m (16 m^2). There were three organic fertilizers treatments, namely poultry manure, wood ash and rice bran, applied at 6 t/ha with a reference treatment NPK 15-15-15 fertilizer applied at 300 kg/ha and a control treatment (no fertilizer). The experiment was arranged in a randomized complete block design and replicated three times. The choice of 6 t/ha for wood ash, rice bran and poultry manure in the experiment was based on the works of Folorunso (1999) and Moyin-Jesu (2008) which carried out extensive research on determination of soil critical levels for N, P, K, Ca and Mg using organic amendments such as rice bran, cocoa husk, poultry manure, wood ash and pig manure for vegetables. They reported that application of these organic fertilizer materials at 6 t/ha was the best critical level for optimum yield of crop in the study area.

The organic amendments were incorporated into the soil using hand trowel one week before transplanting the seedlings. Twenty-day-old seedlings were transplanted to the experimental plots at a spacing of 50 × 50 cm. After transplanting, irrigation was done every mornings and evenings until the rain was steady to allow full establishment.

A low-cost shade made of erected bamboo poles and the top covered with palm fronds was constructed over the transplanted cabbage seedlings to reduce partially direct impact of sun radiations and enhanced the prevailing environment of the site to favour sustainable cabbage production.

Weeding was done manually 10 days after transplanting and continued at 2 weeks interval until harvesting. Cabbage seedlings were sprayed with Avesthrin (Cypermethrin 10 EC) at 10 ml/litre of water at 3 weeks interval until 10 weeks after harvesting (WAT) to control leaf defoliating beetles and caterpillars. Growth parameters such as height, stem girth (cm), leaf area (cm^2) and number of leaves were measured starting from 15 days after transplanting (DAT) until 80 days after transplanting. The leaf area was estimated using the non-destructive and accurate method of Jamal et al. (2009) based on determination of individual leaf area using a predictive equation constructed from leaf length (L) and width (W) and their combination.

Harvest of matured cabbage head started at 82 DAT and continued as they attain maturity in each experimental plot on treatment basis, and yield parameters such as cabbage head weight, cabbage head length and cabbage head girth were measured. In addition, it was observed that new shoots began to grow at the base of the harvested stem.

2.7 Soil analysis after harvesting cabbage

After harvesting, soil samples were taken from each plot using soil auger, bagged, air-dried and sieved for routine laboratory analysis. Soil pH, organic matter (O.M), N, P, K, Ca and Mg were determined as described for the pre-cropping soil analysis.

2.8 Statistical analysis

The average data of 2011 and 2012 cropping seasons on cabbage height, stem girth, leaf area, number of leaves, head weight, head girth and head length were analysed using ANOVA F test, and the treatment means were compared using Duncan multiple range test at 5 % level of significance (Gomez and Gomez (1984).

Table 4. Effect of different organic fertilizers on the growth parameters of cabbage plants

Treatments	Plant height (cm)			Stem girth (cm)			Leaf number			Leaf area (cm^2)		
	2011	2012	Mean	2011	2012	Mean	2011	2012	Mean	2011	2012	Mean
Control	4.00e	4.00e	4.00e	2.94e	2.70a	2.82e	4.64e	4.70e	4.67e	25.73e	26.11e	25.92e
NPK 15-15-15	10.53b	10.75b	10.64b	5.66b	5.30b	5.48b	8.10b	8.16b	8.13b	74.90b	73.60b	74.75b
Rice bran	8.50d	8.30d	8.40d	4.90d	5.18c	5.04d	7.46d	7.48d	7.47d	37.36d	36.20d	36.73d
Wood ash	8.70c	8.90c	8.80c	5.04c	5.14d	5.09c	7.50c	7.56c	7.53c	49.42c	48.30c	48.86c
Poultry manure	12.88a	12.40a	12.69a	6.84a	6.70a	6.77a	8.96a	9.10a	9.03a	88.88a	87.90a	88.39a

Treatment means within each group followed by the same letters are not significantly different from each other using Duncan multiple range test at 5 % level of significance

3. RESULTS

3.1 Effect of organic fertilizers on the growth parameters of cabbage between 15 and 80 days after transplanting (DAT)

There were significant increases ($P < 0.05$) in the plant height, stem girth, number of leaves/plant and leaf area of cabbage under different organic fertilizer treatments compared to the control treatment (Table 4).

The highest values of cabbage plant height, stem girth, number of leaves and leaf area were obtained with poultry manure followed by NPK 15-15-15, wood ash and rice bran, respectively. The application of poultry manure treatment increased cabbage plant height, stem girth, leaf number/plant and leaf area by 17, 19, 10 and 16 % compared to NPK 15-15-15 fertilizer. The cabbage plant height, stem girth, number of leaves/plant and leaf area increased by 34, 26, 18 and 16 % with application of poultry manure compared to rice bran. The least values of growth parameters in cabbage were also obtained under the control treatment.

3.2 Effect of organic fertilizers on the head yield parameters of cabbage plant

There were significant increases ($P < 0.05$) in the head yield parameters (head weight t/ha, head girth and head length) of cabbage under different organic fertilizers compared to the control treatment (Table 5).

The highest values of cabbage head yield parameters were obtained with poultry manure followed by wood ash and rice bran, respectively. When compared to NPK 15-15-15, application of poultry manure also increased cabbage head weight, head girth and head length by 17, 18 and 8 %, respectively.

The application of poultry manure also increased cabbage head weight, head girth and head length by 23, 21 and 13 % compared to the wood ash treatment, respectively. The least values of cabbage yield parameters were obtained under the control treatment.

Table 5. Cabbage head yield parameters under different organic fertilizer treatments

Treatments	Head weight (t/ha)			Head girth (cm)			Head length (cm)		
	2011	2012	Mean	2011	2012	Mean	2011	2012	Mean
Control	0.90e	0.94e	0.92e	12.36e	12.30e	12.33e	9.10e	8.90e	9.0e
NPK 15-15-15	29.30b	28.90b	29.02b	31.80b	32.20b	32.00b	19.20b	18.30b	19.0b
Rice bran	22.60d	22.20d	22.40d	24.87d	24.57d	24.67d	16.80d	17.20d	17.00d
Wood ash	27.20c	27.40c	27.30c	29.90c	30.76c	30.33c	18.40c	17.60c	18.00c
Poultry manure	35.30a	34.90a	35.10a	38.56a	38.10a	38.33a	20.30a	20.70a	20.50a

Treatment means within each group followed by the same letters are not significantly different from each other using Duncan multiple range test at 5 % level of significance

3.3 Soil chemical analysis after harvesting cabbage

Significant increases ($P < 0.05$) occurred in the soil N, P, K, Ca and Mg under different organic fertilizers compared to the control treatment (Table 6).

The soils which received NPK 15-15-15 fertilizer had the highest values of soil N, P, K, and the lowest contents of O.M, Ca and Mg. The highest values of soil organic matter (O.M) and moderate values of soil P, K, Ca and Mg were obtained with the application of poultry manure. The application of wood ash resulted in the highest values of soil pH, Ca and Mg compared to other treatments, while moderate values of soil pH, K, Ca and low % N were obtained with rice bran.

The soil N, P and K increased by 54, 53 and 55 % with application of NPK 15-15-15 compared to rice bran treatment. Also, it had higher K/Ca, K/Mg and P/Mg ratio of 93:1, 74:1 and 572:1 P/Mg, respectively, compared to 2:1 K/Ca, 2:1 K/Mg and 16:1 P/Mg in rice bran treatment.

The soil O.M, % N and P increased by 49, 26 and 39 %, respectively, with application of poultry manure compared to wood ash. However, soil pH, K, Ca and Mg were increased by 3, 30, 20 and 12 %, respectively, under wood ash treatment compared to poultry manure. The least values of soil pH, N, P, K, Ca and Mg nutrients were obtained under control treatment.

4. DISCUSSION

The least values of soil, growth and head yield parameters (number of leaf, leaf area, stem height, head length, head girth and head weight) obtained under the control treatment compared to other treatments could be as a result of initial low soil nutrients status that often characterize continuous cultivation of land without fertilizer application. This finding was supported by Mohammed and Solaiman (2012) who worked on the efficacy of fertilizers on the growth and yield of cabbage and reported that nutrients supply was an important input for realizing higher cabbage yield. Hence, the application of fertilizers, especially organic fertilizers, enhanced both soil and crop productivity in the tropics.

Table 6. Effect of different organic fertilizers on the soil chemical composition after harvesting cabbage

Treatment	Soil pH			O.M (%)			N (%)			P (mg/kg)		
	2011	2012	Mean	2011	2012	Mean	2011	2012	Mean	2011	2012	Mean
Control	5.30d	5.10e	5.20de	0.19e	0.21e	0.20e	0.029e	0.031e	0.03e	3.30e	3.10e	3.20e
NPK 15-15-15	5.12e	5.14d	5.13e	0.27d	0.23d	0.25de	0.39a	0.35a	0.37a	28.40a	28.80a	28.60a
Rice bran	7.16a	7.12a	7.14a	1.20b	1.22b	1.21b	0.25c	0.27c	0.26c	15.90c	15.52c	15.71c
Wood ash	6.60c	6.60c	6.62c	1.15c	1.11c	1.13c	0.15d	0.19d	0.17de	13.71d	13.53d	13.65d
Poultry manure	6.92b	7.06b	6.99b	2.30a	2.40a	2.53a	0.36b	0.34b	0.35b	25.55b	25.47b	25.51b

Treatment	Kmmol/kg			Ca (mmol/kg)			Mg (mmol/kg)		
	2011	2012	Mean	2011	2012	Mean	2011	2012	Mean
Control	0.03e	0.05e	0.04e	0.06d	0.04e	0.05d	0.032e	0.028e	0.03d
NPK 15-15-15	3.74b	6.68a	3.71a	0.03e	0.05d	0.04e	0.048d	0.052d	0.05e
Rice bran	3.90a	3.30b	3.10b	1.54a	1.52a	1.53c	1.28a	1.26a	1.27a
Wood ash	1.85d	1.53d	1.69d	0.93c	0.97c	0.95b	0.80c	0.86c	0.83c
Poultry manure	2.45c	2.65c	2.40c	1.24b	1.22b	1.23a	1.13b	1.11b	1.12b

Treatment means within each group followed by the same letters are not significantly different from each other using Duncan multiple range test at 5 % level of significance

The best cabbage head weight, head length, head girth, number of leaf, leaf area, stem girth and plant height obtained from the application of poultry manure could be attributed to its balanced nutrient contents. The least value of C/N ratio of poultry manure also encouraged faster decomposition and quick release of nutrients for crop uptake and higher head yield parameters. This observation was supported by Ijoyah and Sophie (2009) who reported that the application of poultry manure increased cabbage yield. However, the head yield of 35.10 t/ha of cabbage obtained in this experiment was higher than 25.76 t/ha reported by Ijoyah and Sophie (2009). The prior processing and stacking of poultry manure enhanced quick decomposition and reduction in C/N ratio. This could be responsible for quick nutrient release and uptake by cabbage crop and hence higher cabbage head yield.

The increased soil pH obtained by the application of wood ash was attributed to its high K, Ca and Mg contents which served as base saturation agents and consequently increased soil buffering capacity (Moyin-Jesu and Ogochukwu 2014). The soil pH influenced nutrient availability and uptake as reported by Obatolu (1995) that oil palm bunch ash, wood ash and cocoa pod husk improved soil K, Ca and Mg nutrients and corrected soil acidity in an Alfisol under cultivation of coffee and corn.

The lower values of cabbage growth and head yield parameters under rice bran treatment compared to wood ash and poultry manure could be adduced to the fact that higher C/N ratio (23:3:1) of rice bran compared to poultry manure decelerated the residue decomposition rate and subsequently it had the least values of % N, P, K, Ca and Mg. The observation was supported by Moyin-Jesu (2007) who reported that the growth and yield performance of Okra were very low under the application of rice bran and sawdust. However, the processing of the organic fertilizer (rice bran) by composting was still responsible for reduction of C/N ratio and the results would have been extremely different, if the rice bran was not processed because of its high C/N ratio of 135:1. This was corroborated by Adebayo and Olayinka (1984) who used the unprocessed forms of oil palm bunch ash, sawdust and rice bran to grow maize; hence, prior processing of the organic fertilizers before application is important for good crop yield. Nevertheless, the inclusion of rice bran as source of fertilizer was important because it has been reported by Moyin-Jesu and Ojeniyi (2006) that rice bran, wood ash and spent grain were good in the reduction of soil bulk density and improved % porosity which are good soil fertility indicators (i.e. soil physical properties).

The application of NPK 15-15-15 fertilizer significantly increased the cabbage head yield and growth parameters more than that of rice bran, and this could be traced to the supply of readily available nutrients from the NPK fertilizer to the plant. This observation agreed with that of Makinde (2013) who reported that an increase in the readily available nitrate from the NPK fertilizer unlike the organic manure which must be mineralized before being utilized by crops. Moyin-Jesu (2013a) also reported that nitrogen increased the vegetative and yield performance of garden egg crop.

However, the high content of nitrogen in the NPK fertilizer and its continuous use might be responsible for excessive vegetative growth of cabbage and subsequently delayed maturity. This could explain the lower values of cabbage head yield and growth parameters compared to the poultry manure treatment. Also, Tong et al. (1997) reported higher rate of nitrate nitrogen leaching which polluted underground water particularly with the continuous use of chemical fertilizers compared to the organic fertilizers. Furthermore, Moyin-Jesu (2012) reported that the high K/Ca, K/Mg and P/Mg ratios would cause nutrient imbalance and made availability of nutrients such as K, Ca, and Mg difficult for both immediate and subsequent cropping of cabbage.

5. CONCLUSION

This experiment showed that the use of the various organic fertilizers (poultry manure, wood ash and rice bran) applied at 6 t/ha increased the growth, head yield of cabbage and soil N, P, K, Ca, Mg, pH and organic matter (O.M). It is recommended that poultry manure be applied at 6 t/ha for increased nutrient availability, growth and head yield of cabbage as well as enhancing sustainable cultivation of cabbage on commercial basis. The use of poultry manure would replace the application of large quantity of NPK 15-15-15 fertilizer.

The recommendation is important because the inorganic fertilizers are becoming very expensive to be purchased by small-scale and commercial farmers of cabbage. In addition, these organic fertilizers appear to have beneficial secondary effects on soil properties and ensuring environmental sustainability.

REFERENCES

1. Adebayo A, Olayinka A (1984) The effect of methods of application of sawdust on plant growth, nutrient uptake and soil chemical properties. Ife J Agric 2(1):36–44

2. Afshar RK, Chaichi MR, Assareh MH, Hashemi M, Liaghat A (2014) Interactive effect of deficit irrigation and soil organic amendments on seed yield and flavonolignan production of milk thistle (*Silybum marianum* L, Gaertn). Ind Crop Prod 58:166–172

3. Agboola AA, Corey RB (1973) Soil testing N, P, K for maize in the soils derived from metamorphic and igneous rocks of Western State of Nigeria. J West Afr Sci. Assoc 17:93–100

4. AOAC (1970) Official method of analysis 12th eds. Association of official Analytical Chemists Arlington, VA

5. Awodun MA (2007) Effect of sawdust ash on nutrient status, growth and yieldof Cowpea (*Vigna unguiculata* L. Walp). Asian J Agric Res 1:92–96.

6. Bouycous H (1951) Mechanical analysis of soil using hydrometer method analytical chem. Acta 22:32–34

7. Crockford L, Nowel P (1956) Laboratory manual of physical chemistry. Exp 31 and 32. John Wiley and Sons, New York

8. Folorunso OO (1999) Use of plant residues for improving soil fertility and yield of Okra (*Abelmuschus esculentum* L Moench) and Amaranthus (*Amaranthus viridis* L). PhD thesis, School of Agriculture and Agriculture Technology, Federal University of Technology, Akure, Nigeria: 45–50

9. Folorunso OO, Agboola AA, Adeoye GO (2000) Evaluation of three fertilizer models for P and K recommendation in maize (Zea Mays L). J Tech Educ 2:237–253

10. Gomez KA, Gomez AA (1984) Statistical procedures for agricultural research, 2nd edn. John Wiley and Sons, New York

11. Hague KMF (2006) Yield and nutritional quality of cabbage as affected by nitrogen and phosphorous fertilization. Bangladeshi J Sci Med Res 41:41–46

12. Ijoyah MO, Sophie VL (2009) Effect of different levels of decomposed poultry manure on the yield of cabbage (*Brassica Oleraceae* L). Agro Sci J Trop Agric Food Environ Ext 8(1):20–23

13. Jackson ML (1958) Soil chemical analysis. N.J Prentice Hall, Englewood Cliffs, pp 57–69

14. Jackson ML (1964) Soil chemical analysis. N.J Prentice Hall, Englewood Cliffs, pp 86–92

15. Jamal Ali Olfati, Peyvast G, Samavi M, Salehi M, Mahdipour M, Nosratie-Rad Z (2009) Comparison of leaf area estimation from linear measurement of Red cabbage. Int J Veg Sci 15:185–192.

16. John J (1997) Drug and food interaction. J Nutr Fla 9(2):7–9

17. Kwari JD, Bababe B, Shehu HE (1991) Effect of cow dung on CEC and accumulation of mineral nitrogen, phosphorus and sulphur in soils of Borno state. In: Ojeniyi S.O (eds). Proceedings 19th Annual conference of Soil Sci Soc, Nigeria, Nov 17–21, 1991. University of Maiduguri: 154–158

18. Makinde A (2013) Effect of inorganic fertilizer on the growth and nutrient composition of moringa. J Emerg Trends Eng Appl Sci (JETEAS) 4(2):34–343

19. Moamogwe M (1995) Adaptation trial of introduced cabbage cultivars ARP training reports between 1993–1995. AVRDC Afr Reg Program Arusha Tanzan 4(1):27–29

20. Mohammed RH, Solaiman AHM (2012) Efficacy of organic and inorganic fertilizers on the growth of cabbage (*Brassica Oleraceae* L). Int J Agric Crop Sci 4(3):128–138. ISBN 2227–670X

21. Moyin-Jesu EI (2007) Use of plant residues for improving soil fertility, pod nutrients, root growth and pod weight of Okra *(Abelmuschus esculentum* L). Bioresour Technol 98(2007):2057–2064

22. Moyin-Jesu EI (2008) Determination of soil nutrient levels for maximum yield of Okra using sole and amended plant residues. J Trop Agric Sci 31(2):233–245 **(Malaysia)**

23. Moyin-Jesu EI (2012) Simple and blended organic fertilizers improve soil fertility of degraded nursery soil for production of Kolanut (*Cola acuminate*) seedlings in Nigeria. In: Tech Open Publishers. Chapter book 15. EdsJoann K. Whalen. ISBN: 978-953-307-945-5. p 209–306.

24. Moyin-Jesu EI (2013a) Evaluation of modified neem leaf and wood ash extracts on the soil fertility improvement on growth and yield of garden egg plant. (*Solanium melongena*L.) Colombia International Publishing. Am J Agric Sci Tech 1(3):77–76.

25. Moyin-Jesu EI, Odewande A (2013) Comparative evaluation of different tillage practices and method of planting on soil properties, growth and head yield of cabbage. Am J Res Commun 1:1–20

26. Moyin-Jesu EI, Ogochukwu AI (2014) Comparative evaluation of different organic fertilizer effects on soil fertility, leaf chemical composition and growth performance of coconut (*Cocos nucifera*) seedlings. Int J Plant Soil Sci 3(6):737–750.

27. Moyin-Jesu EI, Ojeniyi SO (2006) Effects of sole and amended plant residues on soil nutrient contents and yield of Okra (*Abelimoschus esculentum* L). Discov Innov J 18(4):318–326

28. Murphy J, Riley JP (1962) A modified single solution method for determination of phosphate in natural waters. Anal Chem Acta 27:31–36

29. Obatolu CR (1995) Nutrient balance sheet of Alfisol grown to coffee and maize using organic fertilizers. In: Agboola AA (eds) Proceedings 3rd Annual Conference of All African Soil Sci Society, University of Ibadan, Ibadan, August 20–23,1995: 250–256

30. Obi ME, Ofonduro CO (1997) The effects of soil amendments on soil physical properties of a severely degraded sandy loam soil in south Eastern Nigeria. Usman Danfodio University, Sokoto, pp 30–35 **(March 2–7)**

31. Ojeniyi SO, Oso OP, Arotolu AA (2001) Response of vegetables to wood ash fertilizer. In: Obigbesan OO (ed) Proceedings 30th annual conference of agric society of nigeria. University of Agriculture, Abeokuta, pp 147–150

32. Sobulo RA, Osiname OA (1981) Soils and fertilizer use in Western Nigeria. Tech Bull no. 11. Institute of Agricultural Research and Training, University of Ife, Nigeria

33. Soil Survey Staff (1999) Soil taxonomy. A basic system for soil classification for making and interpreting soil surveys agric handbook no. 436. USDA, Washington DC

34. Solomon MG, Ogeh JS (1995) Use of leguminous plants and rice husk as fertilizer materials. In: Agboola AA (eds). Proceedings 3rd All African Soil Science Conference August 20–23, 1995, University of Ibadan. 327–333

35. Tindall M (2000) Mineral and organic fertilizing in cabbage and their residual effect for commercial cultivation on yield and quality performance of cabbage. Hort Brass 6:15–20

36. Tong Y, Ove E, Dianquing L, Harald G (1997) Effect of organic manure and chemical fertilizers on nitrogen uptake and nitrate leaching in a Eum orthic anthrosols profile. Nutr Cycl Agroecosyst 48(3):225–229

37. Walkley A, Black IA (1934) An examination of degtajroff method for determining soil organic acid filtration method. Soil Sci 37:29–38

Chapter 10

Effects of Bentonite, Charcoal and Corncob for Soil Improvement and Growth Characteristics of Teak Seedling Planted on Acrisols in Northeast Thailand

Masazumi Kayama [1,*], Suchat Nimpila [2], Sutjaporn Hongthong [2], Reiji Yoneda [1], Wilawan Wichiennopparat [2], Woraphun Himmapan [2], Tosporn Vacharangkura [2] and Iwao Noda [1]

[1]Japan International Research Center for Agricultural Sciences, 1-1 Ohwashi, Tsukuba, Ibaraki 305-8686, Japan

[2]Forest Research and Development Bureau, Royal Forest Department, 61 Phaholyothin Rd., Chatuchak, Bangkok 10900, Thailand

ABSTRACT

When teak (*Tectona grandis* L. f.) is planted on acrisols in Northeast Thailand, its growth is suppressed by low pH, infertility, and low water holding capacity. To examine materials capable of increasing water holding capacity in soil and improving teak growth, we conducted an experiment with teak seedlings. We selected bentonite, charcoal, and corncob and added these materials at a rate of 4% to sandy soil from northeast Thailand. Teak seedlings were potted on these soils and raised from July 2013 to July

2014. We compared growth, photosynthetic rates, leaf water potential, and concentrations of elements in plant organs among bentonite, charcoal, corncob, and no addition (control) treatments. Water content in the soils was increased for the bentonite and charcoal treatments. Teak seedlings potted in these two conditions did not suffer from drought stress. Comparing the growth traits of the teak seedlings, the charcoal treatment produced larger root growth and promoted the uptake of phosphorus and potassium, whereas the bentonite treatment did not show positive effects on growth or nutrient acquisition. In contrast, the corncob treatment decreased water content in the soil, and teak seedling growth was suppressed. We concluded that charcoal was a useful material to improve teak growth in sandy soils.

KEYWORDS

sandy soil; photosynthesis; nutrients; leaf water potential; chlorophyll

1. INTRODUCTION

Teak (*Tectona grandis* L. f) is one of the most valuable timber species in tropical regions, and it has been planted in various countries [1]. The natural distribution of teak is in tropical deciduous forests in India, Myanmar, Thailand, and Laos [1,2]. In natural teak forests, there are obvious rainy and dry seasons, and the leaves of teak fall in the dry season [1,3,4].

Water is one of the most important factors for the growth of teak. The period of the rainy and dry seasons affect the growth of teak, and increasing the length of the dry season shortened the leaf foliation period [4,5]. In terms of its typical morphological traits, teak has large simple leaves (maximum size is over 2,500 cm^2) [6]. Transpiration and stomatal conductance showed high values compared with other tropical woody species [7,8,9]. As a result, teak consumes large amounts of water for its growth. In addition, the leaves of teak are sensitive to drought stress [10]. Below ground, the majority of the roots are concentrated within a soil depth of 60 cm [1,11], and ground water at this depth can readily run dry [12]. On a suitable substance, teak prefers deep, moist, well-drained sandy loam soil [1,13]. In contrast, habitats with waterlogging suppressed the growth of teak [1,14].

Representative soils in Southeast Asia are acrisols (ultisols in USDA soil taxonomy) [15,16]. Acrisols are characterized by low pH, low nutrient content, and aluminum toxicity [15,17,18]. The growth of teak is restricted by nutrient availability, and deficiency of nutrients. In particular, nitrogen, calcium, and phosphorus affect its growth drastically [3,19,20]. The growth of teak is also suppressed by low pH and aluminum toxicity [3,21]. Moreover, clays in acrisols at the surface are leached, and the water holding capacity at the surface is low [16,17]. The growth of teak showed negative correlations with the content of sand, and a plantation with a high content of sand in the soil showed low growth potential for teak [3,22,23]. Acrisols in Northeast Thailand are described as a "light textured sandy soil" (abbreviated to sandy soil), with low content of clay, low pH, low fertility, and low water holding capacity [24,25]. To improve the low pH and poor fertility of acrisol, correction of acidity and fertilization could be applied to accelerate the growth of teak [21,26,27]. However, there has been no research examining how improvements in water holding capacity of acrisols could affect teak growth.

To improve the low water holding capacity of acrisols, some materials have been shown to be effective. In a previous experiment, Ramesh *et al.* [28] added fly ash to planted teak growing on lixisols. The growth of teak was accelerated by the addition of fly ash; however, an obvious increase in water holding capacity was not observed with this treatment. On the other hand, among materials to increase water holding capacity, bentonite is a clay consisting mostly of montmorillonite that has a high water holding capacity [23,29]. In addition, charcoal (biochar) and corncob have a high water holding capacity [30,31,32,33,34]. We considered that these materials may increase the water holding capacity of acrisols, and thereby accelerate the growth of teak. However, previous research did not examine the effects of these materials on the growth of teak.

The aim of our research was to examine the availability of materials to improve suitable water holding capacity and increase teak growth. We conducted a pot experiment with teak seedlings in acrisol with the addition of bentonite, charcoal, or corncob. To verify the effects of these three materials, we examined the ecophysiological traits of teak seedlings, specifically (1) the growth characteristics of seedlings, (2) leaf water potential, (3) photosynthetic rate, and (4) concentrations of elements in plant organs. These parameters were compared with seedlings potted in only acrisol. Our research is a preliminary step to establish silvicultural methods for teak in acrisol in Southeast Asia.

2. MATERIALS AND METHODS

2.1. Study Site

Our experiment was conducted at the Northeast Forest Seed Center located in Khon Kaen Province in northeastern Thailand (16°16′ N, 102°47′ E, 191 m a.s.l.). This center has been measuring meteorological data. The mean annual precipitation was 1104 mm·year^{-1} and annual mean, maximum, and minimum temperatures were 28.3 °C, 40 °C, and 13 °C, respectively (from 2008 to 2012, Northeast Forest Seed Center, unpublished data). Precipitation is concentrated from May to October (Northeast Forest Seed Center, unpublished data). During the experimental period, monthly precipitation and average maximum and minimum temperatures are shown in Figure 1.

Farmers living in this district are interested in the farming of teak [35]. In Khon Kaen Province, we previously published a soil suitability map of teak plantation for farmers [36]. Khon Kaen Province spreads from lowland areas along several rivers (36.9%). These areas suffer waterlogging and are recognized as unsuitable for teak planting. Sandy soil is also distributed over a large area (13.5%). These areas are categorized as moderately suitable areas for teak planting. To accelerate the growth of teak planted in sandy soil, improvement of the water holding capacity is essential. In the case of agriculture in northeast Thailand, soil improvement using clay could increase a yield in crops [37]. We considered that the method of soil improvement will also apply for silviculture.

2.2. Preparation of Materials, Soil, and Teak Seedlings

For the ratio of bentonite, charcoal, and corncob to add to sandy soil, we established 4% as being suitable because the literature showed positive effects at this ratio [26,28,38]. In July 2013, we prepared 7 kg each of bentonite and charcoal (made from *Acacia* sp.) bought from a market in Thailand. For the corncob treatment, we received 7 kg from a farmer in

Khon Kaen Province. Charcoal and corncob were crushed to particle sizes under 5 mm and 1 cm in diameter, respectively. We also collected 600 kg of sandy soil from a land at the Northeast Forest Seed Center. Our experiment was conducted without fertilization because we had to examine the effects of nutrients in bentonite, charcoal, and corncob.

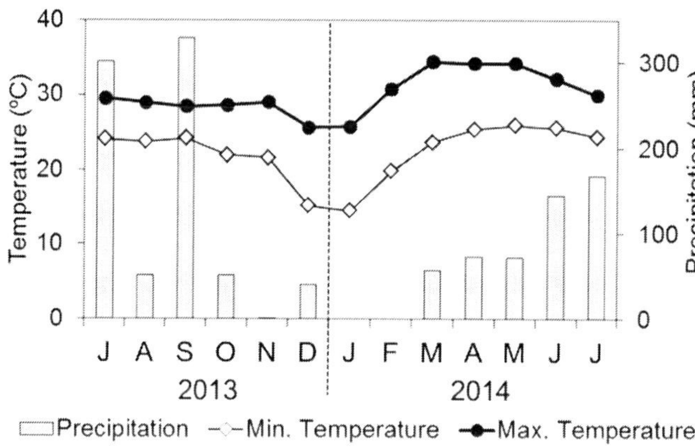

Figure 1. Monthly precipitation and maximum and minimum temperatures at the Northeast Forest Seed Center, Khon Kaen (from July 2013 to July 2014). Note: letters under the horizontal axis are the start of each month. The same format is used in subsequent figures.

The teak seedlings were prepared by a tissue culture technique. The locality of the teak clone was the Province of Mae Hong Son (clone number 21), and this clone has been planted in various places (Royal Forest Department, unpublished data). Teak seedlings were raised from March to June 2013 at the Teak Improvement Center, Lampang, Thailand. In July 2013, 74 teak seedlings were transported to the Northeast Forest Seed Center, Khon Kaen. In addition, we prepared eight tables for raising the teak seedlings.

2.3. Establishment of the Pot Experiment

We established bentonite, charcoal, and corncob treatments by adding them at a rate of 4% into the sandy soil. We mixed the components, and the total weight of the mixed soil was 153 kg per treatment. These mixed soils were put into 17 pots (depth: 22.5 cm, diameter: 27 cm, volume: 8.5 L), and the soil weight per pot was 9 kg. We also prepared 17 pots filled with 9 kg each of sandy soil alone as the control treatment. Before transplantation, 50 g of soil samples were collected from four pots for each treatment.

We potted 64 teak seedlings in 16 of pots for each treatment in July 2013. In each seedling, 0.3 kg of original soil was retained, and this soil was buried in the pot. We also set a soil sensor (ML2, Delta-T Devices Ltd., Cambridge, UK) in one pot per treatment. The water content of the soil was measured continuously. The remaining 10 teak seedlings were sampled at this time. Eight pots (two pots per treatment) were put on the eight tables, and these tables were established in the field in sunny conditions (Figure 2). The four pots containing the soil sensors were also put on the tables (Figure 2). The logger connected to the soil sensor was fixed to a pole made of concrete.

After these works, all pots (including pots with soil sensors) were irrigated with 1 L of water in the morning of every other day. In rainy days, irrigation was not conducted. Irrigation was carried out until the end of our experiment (July 2014).

2.4. Soil Analysis

We measured soil texture and chemical properties including pH and concentrations of C, N, exchangeable P, base cations, Al, and Fe. Four soils of each treatment sampled before transplantation were used. We also collected four samples of original soil from seedlings sampled in July 2013. To determine the pH of the soil, 25 mL of distilled water was added to 10 g fresh soil to make a homogenized mixture [39]. This mixture was then shaken for 1 h and the pH was measured using a pH meter (SG2, Mettler Toledo, Zürich, Switzerland). Prior to chemical analysis, we conducted air-drying of soil samples.

The soil texture was determined by the hydrometer method [40]. The concentrations of C and N in dried soils were determined using a nitrogen and carbon analyzer (Flash 2000, Thermo Fisher Scientific, Waltham, MA, USA). Exchangeable P was separated using dilute acid fluoride [41] by shaking for 1 minute. P in the extracted solution was determined by the molybdenum blue method [42] using a spectrophotometer (U-1800, Shimadzu, Kyoto, Japan). Exchangeable base cations (Ca, Mg, K, Na) and Mn were quantified by mixing 4 g of dry soil with 100 ml of 1 M ammonium acetate solution, and shaking for 1 h [41]. Solutions for exchanged Fe and Al were obtained by adding 20 ml of 1 M KCl to 1 g of dry soil and shaking for 1 h [41]. Base cations in the extracted solutions were analyzed using an atomic absorption spectrophotometer (AAnalyst 300, Perkin-Elmer, Norwalk, CT). For Al and Fe, the extracted solution was analyzed using an inductivity coupled plasma (ICP) analyzer (ICPE-9000, Shimadzu, Kyoto, Japan).

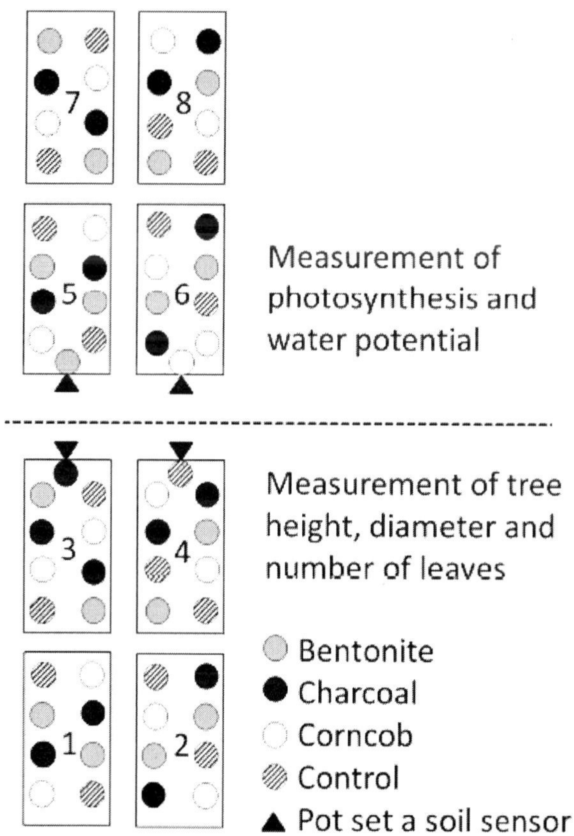

Figure 2. Schematic of experimental treatment design, eight seedlings (two pots per treatment) were put on a table.

2.5. Measurement of Teak Seedlings

For the measurement of teak seedlings, eight seedlings for each treatment were used for growth measurements (seedlings put on No.1 to 4 tables, Figure 2). We measured tree height and basal diameter at six time points (July 2013, October 2013, December 2013, February 2014, April 2014, and July 2014). We also measured length and width for all leaves of the teak seedlings. The leaf area of teak seedling was calculated from the equation of Tondjo *et al.* [43], as follows:

$$A = 0.60 \times L \times \mathrm{Wi}$$

where A is the leaf area, L is the leaf length, and Wi is the leaf width.

From September 2013, some teak seedlings elongated new shoots from the base of the root. Leaves foliated from new shoots were called "low position leaves", and separated from other leaves, which were called "high position leaves".

From our observations, teak leaves foliated in response to continuous irrigation. To estimate leaf productivity and turnover, the numbers of leaves of seedlings were counted every week for eight individuals in each treatment. We calculated the leaf longevity from the dates of occurrence and defoliation of leaves. In addition, we counted the total number of leaves produced in the experimental period.

2.6. Measurement of Photosynthetic Rate

We measured area-based photosynthetic rate at light saturation (P_{sat}) and stomatal conductance (gs) for teak leaves located second from the top. For the measurement of photosynthetic rate and leaf water potential, the eight teak seedlings not used for the measurement of growth parameters were used by treatment (seedlings put on No. 5 to 8 tables, Figure 2). There was no irrigation to eliminate the effects of water in a pot on the day when we conducted these measurements.

When we measured the photosynthetic rate, the leaves of some seedlings were immature or senescent. To provide uniformly mature teak leaves, we selected four teak leaves among eight seedlings. We measured P_{sat} five times (October 2013, December 2013, February 2014, April 2014, and July 2014), and the time of measurement was uniformly from 09:00 to 11:00 (UTC+7).

Measurements were made using a portable gas analyzer (LI-6400, LiCor, Lincoln, NE, USA) under steady-state conditions, an ambient temperature of 27 °C, and ambient CO_2 concentration of 38.0 Pa. The LED light source was adjusted to a saturation light level of 1800 μmol $m^{-2}\cdot s^{-1}$ PPF.

2.7. Measurement of Leaf Water Potential, Concentration of Chlorophyll, and Leaf Mass Per Area

After measurement of photosynthetic rate, we measured the leaf water potential of teak leaves five times (October 2013, December 2013, February 2014, April 2014, and July 2014). The leaf used to measure water potential was the same as that used to measure photosynthetic rate. In general, leaf water potential showed the lowest value in the afternoon by transpiration, and showed the highest value by recovery of water during the night [44]. We measured leaf water potential in the afternoon and predawn. Four teak leaves for each treatment were sampled at 1–2 p.m. and 5:30–6:00 a.m. the next day (UTC+7). Sampled shoots were put in a plastic bag that contained a wet filter paper, and kept in a refrigerator. We measured leaf water potential using a pressure chamber (Model 600, PMS Instrument Co., Albany, OR, USA).

After measuring leaf water potential, we analyzed the concentration of nitrogen, chlorophyll, and leaf mass per area (LMA). Photosynthetic rate is closely related to concentrations of nitrogen [44,45,46], chlorophyll [47,48], and leaf thickness [49,50]. We sampled 0.66 cm^2 of teak leaf samples in a freezer. Chlorophyll was extracted from the leaves using dimethyl sulfoxide (DMSO), and then measured using a spectrophotometer. The amounts of chlorophylls a and b were calculated in accordance with the previously reported equation [51]. The remaining leaf samples were oven-dried at 70 °C for three days, and LMA ($g\cdot m^{-2}$) measured by the method of Larcher [44]. Leaf samples were ground to a fine powder using a sample mill (WB-1; Osaka Chemical Co., Osaka, Japan). The concentration of N was determined using an NC analyzer (Sumigraph NC-220F, Sumika Chemical Analysis Service, Tokyo, Japan). The results of the concentration of N were calculated from the area based on N (N_{area}) from the data of LMA.

2.8. Analysis of Biomass and Element Concentrations of Teak Seedlings

To determine the biomass of teak seedlings, we measured the dry mass of leaves, stems and branches, and roots. In July 2014, eight teak seedlings from each treatment used for growth measurements were sampled (seedlings put on No. 1 to 4 tables, Figure 2). Roots of seedlings were washed twice with water to remove soil. The washed seedlings were divided into leaves, stems and branches, and roots. Each component was put into its own envelope and oven-dried at 70 °C for three days. After drying, the dry mass of each component was determined. We also examined the root/shoot ratio as an indicator of teak growth [10,19,21].

Of the elements in the plant organs, we measured the mass base of concentrations of N, P, K, Ca, Mg, Al, and Fe in leaves and roots. Dried samples were ground to a fine powder, and N concentration was determined using the NC analyzer. The remaining samples were digested by the HNO_3-HCl-H_2O_2 method [52]. Concentrations of K, Ca, Mg, Al, and Fe were analyzed using an ICP analyzer. During analysis, we analyzed standard solutions of each element after 30 samples to verify the reliability of the analyses. Concentration of P was determined by the molybdenum blue method using a spectrophotometer (UV-2500PC, Shimadzu, Kyoto, Japan).

2.9. Statistical Analysis

Significant pairwise differences for each variable were tested by Tukey's test using Stat View 5.0 (SAS Institute Inc., Cary, NC, USA). Comparisons were made among the bentonite, charcoal, corncob, and control treatments.

3. RESULTS

3.1. Climate during the Experimental Period

Precipitation in July and September 2013 was high compared with other months (Figure 1). However, precipitation in August 2013 was low, and precipitation was not stable. Precipitation decreased from October 2013, and there was no rain in January or February 2014. From March 2014, precipitation increased gradually.

The maximum temperature was around 29 °C, and this value was stable from July 2013 to November 2013. In December 2013 and January 2014, the maximum temperature decreased below 26 °C, then from February 2014, the maximum temperature increased. The maximum temperature reached the highest value in March 2014 (34.4 °C). The minimum temperature was around 23 °C from July 2013 to November 2013. However, the minimum temperature decreased to 15 °C in December 2013 and January 2014. In March 2014, minimum temperature increased to 25 °C, and this value was stable until July 2014.

3.2. Water Contents in Soil

To examine the minimum value of water content in soil for each season, we collected the lowest water contents before irrigation and rainfall. Averages of these values for each month are shown in Figure 3. The water content was high for the bentonite treatment throughout the whole period. For the charcoal treatment, water content increased from September 2013. In contrast, the corncob treatment showed low water content compared with the other three treatments. Water content in the corncob treatment increased and approached the value of the control treatment from January 2014. Comparing months, the water contents in October and November 2013 were lower than those in other months. From December 2013, water content increased for each treatment, and water content also decreased in April 2014 for each treatment.

Water contents for each treatment were increased by rainfall in September 2013. The maximum values of water contents for each treatment were 0.59 $m^3 \cdot m^{-3}$ for bentonite, 0.41 $m^3 \cdot m^{-3}$ for charcoal, 0.27 $m^3 \cdot m^{-3}$ for corncob, and 0.30 $m^3 \cdot m^{-3}$ for control.

3.3. Soil Properties

For soil texture, the content of sand was over 88.0% for each treatment (Table 1). The contents of silt and clay were low for each treatment. However, the clay content of the bentonite treatment was significantly higher than those for other treatments ($p < 0.05$). Clay content for bentonite alone was 66%.

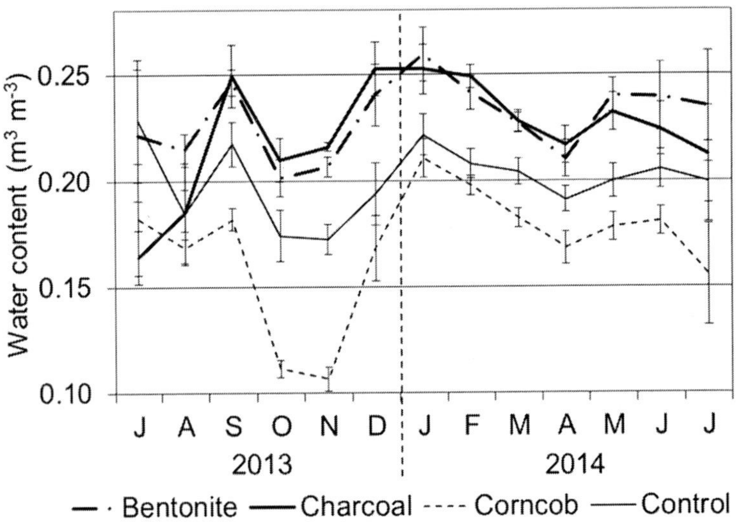

Figure 3. Average of soil water content among the four treatments before irrigation or rainfall (from July 2013 to July 2014, Mean ± SE). Note: the values of water content were averaged among several dates with low values before irrigation.

The pH value was significantly higher for the charcoal treatment than those of other treatments ($p < 0.05$). The concentrations of C, N, Ca, and Mg for the charcoal treatment were also significantly higher than those for the control treatment ($p < 0.05$). In contrast, bentonite treatment resulted in significantly higher concentrations of Mg and Na compared with the control treatment ($p < 0.05$). Fe concentrations with the bentonite treatment were significantly lower than with the control treatment. In the corncob treatment, the concentrations of elements did not show a significant trend.

There was no significant difference in the concentrations of P and Al among the four treatments.

The original soil contained in the seedling pots showed high contents of silt and clay compared with the sandy soil used for each treatment. Concentrations of C, N, Ca, and Mg showed higher values than sandy soil.

3.4. Growth Characteristics

The tree height with the corncob treatment was significantly lower than that for the control treatment in October 2013 and February 2014 (Figure 4, $p < 0.05$). The basal diameter was also significantly smaller for the corncob treatment than the other treatments from October 2013 to February 2014 ($p < 0.05$). In February 2014, some of the original shoots of seedlings in the corncob treatment had died, and as a result the tree height decreased compared with December 2013. However, these seedlings elongated new shoots grown from the base of the roots. From April 2014, seedlings in the corncob treatment accelerated their growth, and the tree height and basal diameter did not show a significant difference among the four treatments. For growth in the other treatments, charcoal treatment showed the highest value of basal diameter in December 2013 ($p < 0.05$); however, this trend did not show from February 2014.

3.5. Leaf Dynamics of Teak Seedling

The total leaf area in October 2013 was significantly larger for the charcoal treatment than those of other treatments (Figure 5, $p < 0.001$). In contrast, the corncob treatment showed small total leaf areas compared with other treatments ($p < 0.001$). However, the total leaf area of the charcoal treatment decreased in December 2013. From this month, the total leaf area did not show a significant difference among the four treatments. For the leaf area at high positions, bentonite treatment resulted in significantly larger areas than those of other treatments in December 2013 and February 2014 ($p < 0.05$). In the control treatment, the leaf area at the high position was decreased in December 2013 ($p < 0.05$) and showed a small trend after this month.

Table 1. Texture and chemical properties on soils of four treatments and original soil with the seedlings (Mean ± SE, $n = 4$). Different letters indicate significant differences among the four treatments calculated by Tukey's test ($p < 0.05$).

Treatment	Texture (%)							
	Sand		Silt		Clay		pH	
Bentonite	89.5 ± 0.1	b	4.5 ± 1.0	a	6.0 ± 0.0	a	5.03 ± 0.11	c
Charcoal	91.0 ± 0.1	ab	5.5 ± 1.9	a	3.5 ± 1.0	b	6.23 ± 0.07	a
Corncob[1]	88.5 ± 0.1	bc	4.3 ± 1.0	a	3.4 ± 1.0	b	5.47 ± 0.10	b
Control	92.0 ± 0.0	a	4.0 ± 0.0	a	4.0 ± 0.0	b	5.24 ± 0.06	bc
Original	75.7 ± 0.2		15.3 ± 0.1		9.0 ± 0.0		5.92 ± 0.16	
	C (mol·kg^{-1})		N (mmol·kg^{-1})		P (mmol·kg^{-1})		Ca (mmol·kg^{-1})	
Bentonite	0.34 ± 0.04	b	7 ± 0	b	0.267 ± 0.127	a	3.3 ± 0.2	b
Charcoal	1.57 ± 0.17	a	31 ± 3	a	0.343 ± 0.086	a	10.7 ± 1.1	a
Corncob	0.65 ± 0.27	b	7 ± 0	b	0.102 ± 0.035	a	1.9 ± 0.0	b
Control	0.54 ± 0.21	b	6 ± 2	b	0.173 ± 0.081	a	2.4 ± 0.3	b
Original	2.82 ± 1.18		178 ± 46		0.406 ± 0.066		18.2 ± 2.9	
	Mg (mmol·kg^{-1})		K (mmol·kg^{-1})		Na (mmol·kg^{-1})		Al (mmol·kg^{-1})	
Bentonite	2.0 ± 0.1	a	1.03 ± 0.08	b	0.926 ± 0.091	a	0.150 ± 0.010	a
Charcoal	1.8 ± 0.1	a	3.02 ± 0.19	a	0.081 ± 0.023	b	0.159 ± 0.003	a
Corncob	1.3 ± 0.0	b	2.79 ± 0.88	ab	0.012 ± 0.012	b	0.145 ± 0.008	a
Control	1.3 ± 0.0	b	1.14 ± 0.06	ab	0.083 ± 0.064	b	0.147 ± 0.010	a
Original	11.8 ± 1.1		2.48 ± 0.20		0.651 ± 0.096		0.147 ± 0.003	
	Fe (μmol·kg^{-1})							
Charcoal	9.4 ± 1.1	ab						
Bentonite	9.0 ± 0.7	b						
Corncob	10.9 ± 1.5	ab						
Control	14.0 ± 0.9	a						
Original	12.0 ± 0.6							

Note: 1: For texture in the "Corncob" treatment, corncob (the particle was larger than sand) contained 3.8% ± 1.0%.

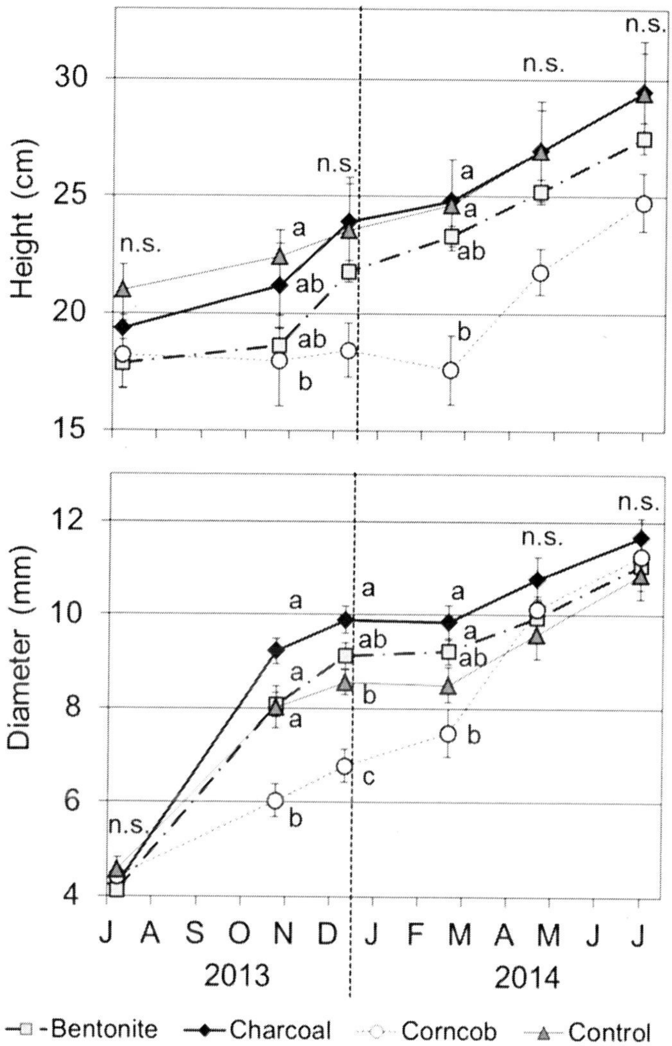

Figure 4. Tree height and basal diameter of teak seedlings grown in four treatments (Mean ± SE, n = 8). Note: divisions on horizontal axis over the letters are the first day of each month. The same format is used in subsequent figures.

The average leaf longevity at high positions was 116 days for bentonite, 90 days for charcoal, 115 days for corncob, and 76 days for control treatments. Leaf longevity was significantly longer for the bentonite and corncob treatments than for the control treatment (Tukey's test, $p < 0.01$). Comparing seasons, the trend in leaf longevity was individually specific and did not show a definite trend for each treatment. During this experiment, the total numbers of leaves produced were 27.6 for bentonite, 28.1 for charcoal, 26.8 for corncob, and 37.4 for control. There was no significant difference in leaf production among the four treatments.

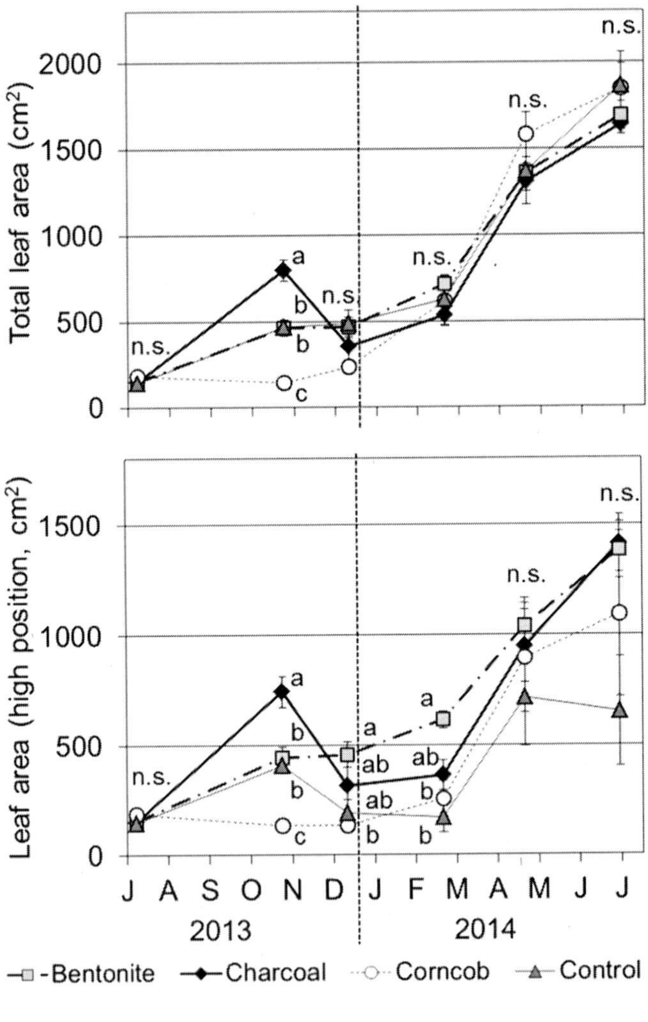

Figure 5. Total and high position (located at the top of stem) leaf areas for teak seedlings grown in the four treatments (Mean ± SE, $n = 8$).

3.6. Photosynthetic Rate

The P_{sat} was significantly higher for bentonite and charcoal treatments than for the corncob treatment at each time point (Figure 6, $p < 0.05$). The P_{sat} for the control treatment decreased drastically from April 2014. Comparing the five time points, P_{sat} in February 2014 was highest for bentonite, charcoal, and control treatments. In the corncob treatment, P_{sat} was highest in October 2013.

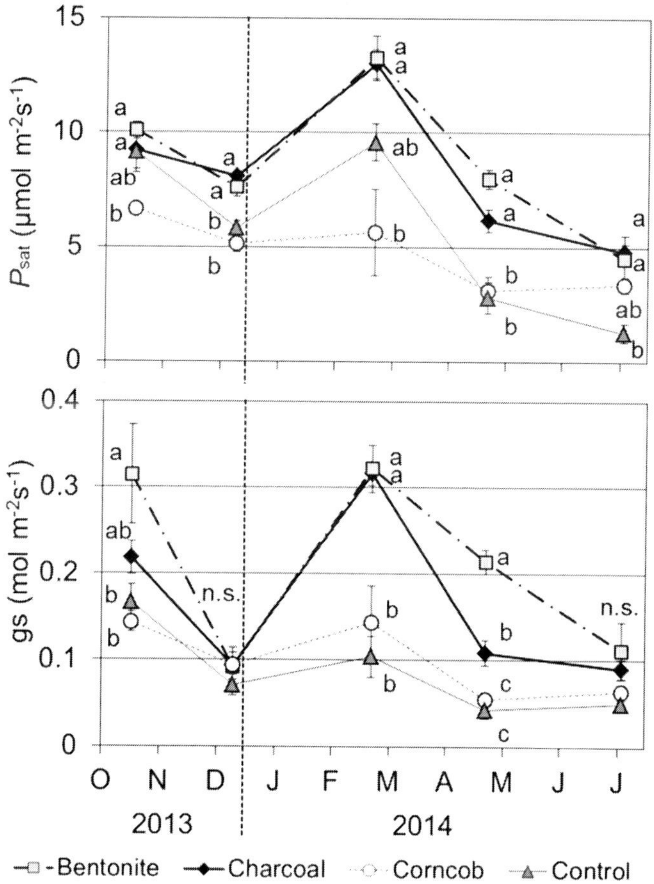

Figure 6. Photosynthetic rate at light saturation (P_{sat}) and stomatal conductance (gs) for teak seedlings grown in the four treatments (9–11 a.m., Mean ± SE, $n = 4$).

The stomatal conductance (gs) with the bentonite treatment was significantly higher than those for the corncob and control treatments in October 2013 and February and April 2014. In the charcoal treatment, gs in February was also high compared with those for the corncob and control treatments. The value of gs decreased in December 2013 and April 2014.

3.7. Leaf Water Potential

Leaf water potential in the afternoon was highest in December 2013 (Figure 7). In contrast, leaf water potential decreased in April 2014, except for the control treatment. Comparing the four treatments, leaf water potential in December 2013 was significantly lower for the corncob treatment than those for the other treatments ($p < 0.05$). At other time points, leaf water potential did not show a significant difference among the four treatments.

The predawn leaf water potential was −0.12 MPa in October and December 2013. The values of leaf water potential decreased in 2014, and in February and April 2014, the value was significantly lower for the control treatment than that for the bentonite treatment ($p < 0.05$).

3.8. Concentrations of Nitrogen and Chlorophyll, and LMA

N_{area} showed the highest value in February 2014; however, N_{area} decreased from April 2014 (Figure 8). Compared the treatments, N_{area} was significantly lower for the charcoal treatment than for the control treatment in October 2013 ($p < 0.05$). At other time points, there were no significant differences among the four treatments.

The concentration of total chlorophyll was also highest in February 2014 for the bentonite, charcoal, and control treatments (Figure 8). From April 2014, the concentrations of chlorophyll decreased for these three treatments. In the case of the corncob treatment, chlorophyll decreased from February 2014, and the value was significantly lower than the control treatment ($p < 0.05$). In October and December 2013, the concentrations of chlorophyll were significantly lower for the charcoal treatment than the control treatment ($p < 0.05$).

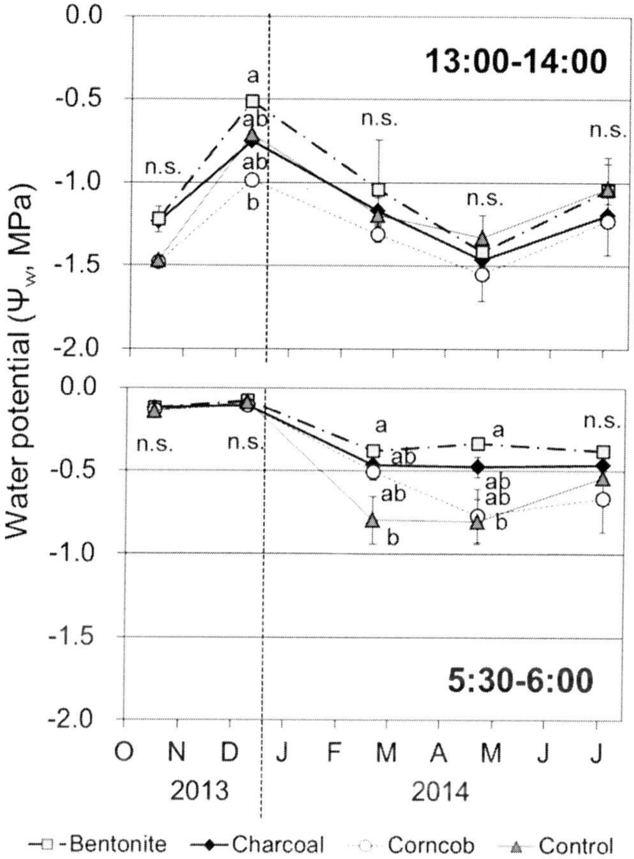

Figure 7. Leaf water potential in the afternoon (1–2 p.m.) and predawn (5:30–6:00 a.m.) for teak seedlings grown in the four treatments (Mean ± SE, *n* = 4).

LMA was highest in July 2014 (Figure 9). In the bentonite and charcoal treatments, LMA was 60 to 70 g·m^{-2} until April 2015. In contrast, the corncob and control treatments showed decreased LMA in February 2014. Comparing the four treatments, LMA in February 2014 was significantly lower for the corncob treatment than that for the bentonite treatment (p< 0.05).

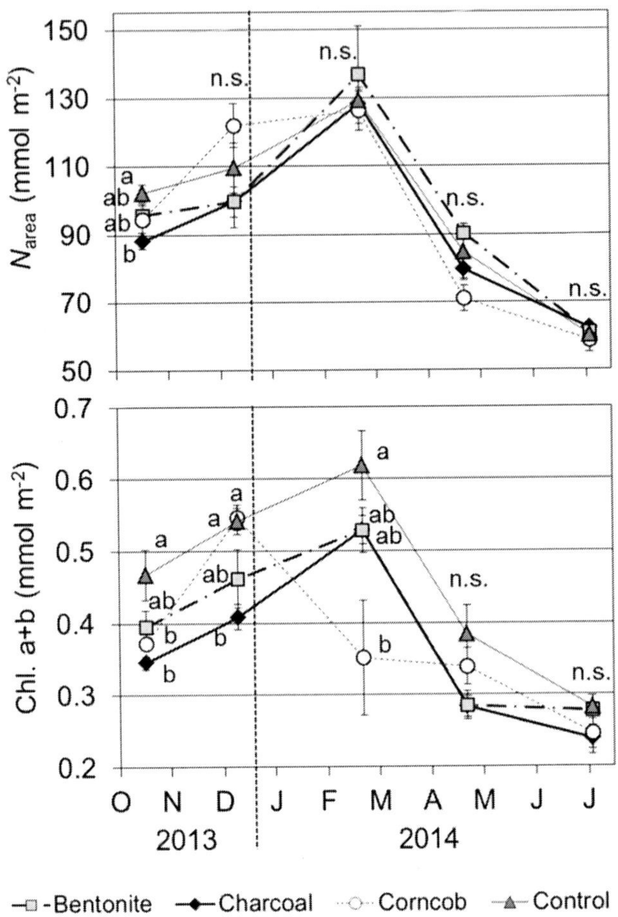

Figure 8. Concentrations of area-based nitrogen (N_{area}) and total chlorophyll (a + b) for teak seedlings grown in the four treatments (9–11 a.m., Mean ± SE, n = 4).

3.9. Biomass of Teak Seedling

The dry masses of each organ in July 2014 increased for each treatment compared with those values in July 2013 (Table 2). For the leaf dry mass, there were no significant differences among the four treatments. The dry mass of stems and branches was significantly larger for the charcoal treatment than for the corncob treatment ($p < 0.01$). In addition, charcoal treatment resulted in significantly higher root dry mass compared with the other treatments ($p < 0.05$). In contrast, root dry mass for the corncob treatment was significantly smaller than the other treatments ($p < 0.001$). The root:shoot ratio was significantly lower for the corncob treatment compared with the other treatments ($p < 0.05$).

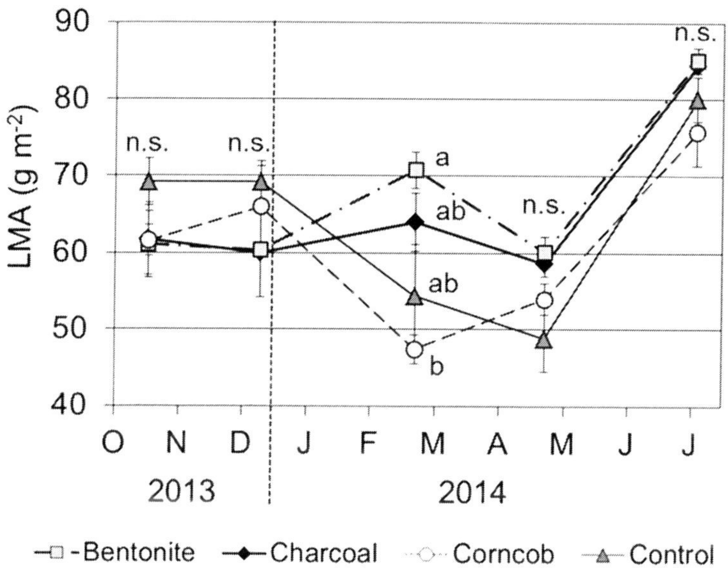

Figure 9. Leaf mass per area (LMA) for teak seedlings grown in the four treatments (Mean ± SE, $n = 4$).

3.10. Concentration of Elements in Plant Organs

The concentrations of P in the leaves of teak seedlings were significantly higher for the charcoal and corncob treatments compared with the control treatment (Table 3, $p < 0.05$). In the charcoal treatment, the concentration of K was also significantly higher compared with the control treatment. For other elements in leaves, there were no significant differences among the four treatments ($p < 0.05$).

Table 2. Dry mass of each organ (leaf, stem and branch, and root) and root:shoot ratio for teak seedlings before transplantation (Mean ± SE, n = 10) and the end of the experiment (July 2014) grown with four treatments (Mean ± SE, n = 8). Different letters indicate significant differences among the four treatments calculated by Tukey's test (p< 0.05).

	Leaf		Stem and Branch		Root		Root:Shoot Ratio	
July 2013	0.9 ± 0.1		0.9 ± 0.1		2.5 ± 0.2		1.48 ± 0.14	
July 2014								
Bentonite	15.4 ± 0.6	a	6.9 ± 0.2	ab	24.4 ± 0.9	b	1.10 ± 0.06	a
Charcoal	15.7 ± 0.6	a	9.0 ± 0.9	a	29.2 ± 1.2	a	1.19 ± 0.03	a
Corncob	17.5 ± 1.5	a	6.2 ± 0.3	b	12.8 ± 1.3	c	0.58 ± 0.10	b
Control	17.4 ± 1.6	a	7.3 ± 0.6	ab	21.9 ± 1.1	b	0.97 ± 1.14	a

Table 3. Concentrations of elements (N, P, K, Ca, Mg, Al, and Fe; μmol·g^{-1} DM) in leaves and roots of teak seedlings grown in four treatments (Mean ± SE, *n* = 8).

Elements		Leaf		Root	
N	Bentonite	746 ± 33	a	624 ± 23	a
	Charcoal	723 ± 40	a	695 ± 38	a
	Corncob	788 ± 42	a	633 ± 24	a
	Control	764 ± 25	a	729 ± 23	a
P	Bentonite	272 ± 18	ab	170 ± 8	b
	Charcoal	351 ± 22	a	209 ± 13	a
	Corncob	324 ± 36	a	185 ± 12	ab
	Control	209 ± 15	b	179 ± 4	ab
K	Bentonite	243 ± 19	ab	193 ± 10	bc
	Charcoal	265 ± 21	a	222 ± 16	ab
	Corncob	235 ± 21	ab	266 ± 14	a
	Control	187 ± 16	b	154 ± 12	c
Ca	Bentonite	174 ± 12	a	332 ± 20	a
	Charcoal	177 ± 12	a	379 ± 48	a
	Corncob	174 ± 19	a	355 ± 30	a
	Control	187 ± 11	a	372 ± 18	a
Mg	Bentonite	111 ± 5	a	166 ± 8	ab
	Charcoal	123 ± 7	a	146 ± 9	b
	Corncob	131 ± 15	a	168 ± 13	ab
	Control	152 ± 13	a	201 ± 13	a
Al	Bentonite	2.52 ± 1.97	a	25.5 ± 4.2	a
	Charcoal	6.33 ± 2.26	a	29.1 ± 4.5	a
	Corncob	3.34 ± 1.45	a	2.2 ± 0.9	b
	Control	0.77 ± 0.62	a	18.0 ± 3.7	a
Fe	Bentonite	2.18 ± 0.14	a	49.5 ± 4.9	a
	Charcoal	2.60 ± 0.36	a	39.2 ± 5.3	ab
	Corncob	2.39 ± 0.16	a	21.4 ± 2.7	b
	Control	2.36 ± 0.51	a	38.0 ± 6.1	ab

Different letters indicate significant differences among the four treatments calculated by Tukey's test ($p < 0.05$).

In roots, the concentration of K was significantly higher for the charcoal and corncob treatments than the control treatment ($p < 0.01$). In the control treatment, the concentration of Mg was high, and significantly higher than for the charcoal treatment ($p < 0.01$). In contrast, the concentration of P in the bentonite treatment was significantly lower than for the charcoal treatment ($p < 0.05$). In the corncob treatment, the concentrations of Al and Fe were low compared with the other treatments.

4. DISCUSSION

Based on the results of the water content before irrigation, the bentonite and charcoal treatments showed high values (Figure 3). These materials probably have the ability to increase water holding capacity. In the bentonite treatment, the clay content was increased by 2% compared with the control treatment (Table 1). Compared with other clay minerals, bentonite has a high liquid limit and high expansibility as a result of water absorbance [53]. We considered that the slight addition of bentonite clay probably absorbed more water, and as a result the water content in the soil increased. This trend was obvious from May 2014 (Figure 3), when precipitation was increased (Figure 1). In the case of charcoal, we could not confirm an obvious change in soil texture (Table 1). Charcoal contains vast numbers of pores, and water can be stored in these pores [30,31,54]. In the charcoal treatment, pores in charcoal can store water, thereby increasing water content in the soil. In corncob, water content was low in October and November 2013, but at its highest in January 2014 (Figure 3). This change in water content may be attributed to decomposition of the corncobs. In general, the addition of organic amendments to soil shows an increase of water holding capacity, porosity and surface area [55]. However, the increase of water holding capacity of the corncob was probably inferior to charcoal and bentonite based on the results of the soil water content (Figure 3).

Compared with the bentonite and charcoal treatments, trends in P_{sat} and leaf water potential in the afternoon and predawn were similar (Figure 6 and Figure 7). Teak seedlings of these treatments could recover water during the night. Thus, teak seedlings grown in the bentonite and charcoal treatments did not show the symptom of water deficiency. However, teak seedlings in the bentonite treatment did not show large values of tree height, basal diameter, or biomass of plant organs (Figure 4, Table 2). Although the teak seedlings did not suffer from drought stress, this was not reflected in obvious growth acceleration.

Meanwhile, the charcoal treatment showed large root biomass in July 2014 (Table 2). Of the differences between bentonite and charcoal treatments, the pH value and the concentrations of N, Ca, and K were higher for the charcoal treatment than those for the bentonite treatment (Table 1). However, the concentrations of N, Ca, and K in plant organs did not show significant differences between the bentonite and charcoal treatments (Figure 8, Table 3). Thus, soil pH and uptakes of N, Ca, and K are not the main factors that affect the differences of root growth. For other nutrient parameters, the concentration of P in roots was significantly higher in the charcoal treatment than in the bentonite treatment (Table 3). However, the concentration of P in soils showed no significant difference between the charcoal and bentonite treatments (Table 1). In the case of teak, uptake of P is related to symbiosis with arbuscular mycorrhizal fungi, and their inoculation accelerates the uptake of P [56,57]. The presence of charcoal promotes infection by arbuscular mycorrhizal fungi [58,59] and root growth in woody species [60,61]. We consider that the presence of charcoal accelerated root growth in teak seedlings. In addition, there is a possibility that teak seedlings in the charcoal treatment may have increased mycorrhizal colonization, and as a result the uptake of P was accelerated.

Compared with bentonite and charcoal, the corncob treatment showed negative effects on the growth of teak. Tree height, basal diameter, and leaf production were suppressed for teak seedlings in the corncob treatment until February 2014 (Figure 4 and Figure 5). In addition, root dry mass of the teak seedlings in July 2014 was smaller for the corncob treatment than those of other treatments (Table 2). One of the causes was considered to be low photosynthetic productivity; P_{sat} and gs were both low up to April 2014 (Figure 6). Soil water content in the corncob treatment was lower than in other treatments (Figure 3), and as a result the stomata of teak leaves were easy to close to prevent water loss by transpiration. The photosynthetic rate was probably suppressed by decreased CO_2 uptake caused by the closed stomata. Moreover, the concentration of chlorophyll in leaves in the corncob treatment was low in February 2014 (Figure 8). In general, chlorophyll concentration is decreased by strong photoinhibition [62,63], and photoinhibition occurs readily in thin leaves [64]. The thin leaves in the corncob treatment in February 2014 (Figure 9) probably suffered from photoinhibition due to strong sunshine; as a result, the concentration of chlorophyll decreased.

Growth in the corncob treatment recovered from April 2014 (Figure 4 and Figure 5). In seedlings in the corncob treatment at the end of experiment, leaf dry mass was not significantly lower than those of the other treatments (Table 2). It seems that supply of nutrients was probably affected by the increase in leaf dry mass. Indeed, the concentrations of P in leaves and K in roots were higher than those of the control treatment (Table 3). Corncob contains little in terms of nutrients [37], and these nutrients may be available only upon decomposition of the corncob. Moreover, the concentrations of Al and Fe in the roots in the corncob treatment were lower than those of the other treatments (Table 3). A previous study showed that the addition of corncob had the capacity to suppress the uptake of Al and accelerate the uptake of P [37]. Recovery of growth in teak seedlings in the corncob treatment may be concerned with an improvement in nutrient uptake due to multiple factors.

Regarding the growth traits of teak seedlings in the control treatment, the predawn leaf water potential of leaves was low (Figure 7), indicating that teak leaves in the control treatment could not recover water during the night.

Comparing our results with the literature, when the growth of teak seedlings was suppressed drastically, the value of the root:shoot ratio increased over 1.0 [10,19,21,65]. In the present study, the root:shoot ratios were over 1.0 for the bentonite and charcoal treatments (Table 2). We considered that teak seedlings in bentonite and charcoal suffered suppressed growth, and one of the causes of this was N deficiency. The N_{area} of each treatment decreased from April 2014 (Figure 8) when leaf production increased (Figure 5). The concentration of N in leaves at the end of the experiment (average, 755 $\mu mol \cdot g^{-1}$) indicated N deficiency (<857 $\mu mol \cdot g^{-1}$) [14]. In contrast, the concentration of P in leaves at this time (average, 289 $\mu mol \cdot g^{-1}$) was much higher and did not indicate any deficiency (<32 $\mu mol \cdot g^{-1}$) [14]. Thus, sandy soil was considered as a rather infertile environment, especially for N. In addition, P_{sat}, gs, and chlorophyll concentrations were also lower from April 2014 (Figure 6 and Figure 8). The photosynthetic rate and chlorophyll concentration show positive correlations with the concentration of N [19,43,44,45,66], and these decreases were related to decreased N in leaves. In terms of stomatal conductance, a leaf suffering from N deficiency can readily suffer from a decrease in leaf water potential [67]. Thus, N deficiency negatively affects physiological parameters in teak seedlings. Meanwhile, the concentration of N in leaves was a high value until February 2014 (Figure 8). The original soil contained with each seedling had a high concentration of N (Table 1), and it seems that the teak seedlings could absorb N from the original soil, and its supply was enough until February 2014.

Despite the N deficiency, we confirmed that teak seedlings were able to produce large numbers of leaves from April to July 2014 (Figure 5). In teak, Ca is also important for leaf production, and Ca deficiency suppresses leaf dry mass [20,21]. Our results showed that the concentration of Ca in leaves (average, 178 $\mu mol \cdot g^{-1}$) was higher than that in the deficiency level (<138 μmol·g^{-1}) [14]. Teak seedlings potted on sandy soil probably accelerated leaf production through the supply of Ca.

In terms of specific growth traits of teak, leaf longevity was quite short (average, 99 days) and leaf turnover was high. Leaf longevity in tropical deciduous tree is three to 12 months [68,69], and the teak seedling is categorized as a species with short leaf longevity. In addition, P_{sat} and gs in December 2013 were low despite the high N_{area} (Figure 6 and Figure 8). The date of measurement of P_{sat} showed that the minimum temperature was 11 °C in the morning. In general, teak shows a range of optimum temperature between 15 and 40 °C [2]. In the case of the rubber tree, chilling stress was suffered below 10 °C, andP_{sat} and gs decreased at this temperature [70]. There is a possibility that the leaves of the teak seedlings in December 2013 were suffering from chilling stress, and as a result, P_{sat} and gs could not recover in the morning.

5. CONCLUSIONS

When teak seedlings were potted in sandy soil mixed with charcoal, root growth was accelerated despite the absence of fertilization. The water content of the soil was also increased, and the teak seedlings did not show the symptom of water deficiency. Moreover, additional charcoal could increase soil pH and nutrients such as N and Ca. Thus, we concluded that charcoal is a useful material to improve the seedling stage of teak growth in sandy soils in northeast Thailand. The application of charcoal for silviculture has been reported in Laos, and several tree species showed accelerated growth from its addition [71]. To establish silviculture methods using charcoal, further verification tests with teak should be conducted in the field with charcoal mixed with sandy soil.

Bentonite also showed a capacity to increase the water content of soil. Teak seedlings potted on sandy soil with bentonite did not suffer from drought stress, but they did not show accelerated growth. One of the reasons for the lack of growth acceleration was the absence of nutrients in bentonite. To apply this treatment for forestry in northeast Thailand, a combination of bentonite and fertilizer would be needed. Bentonite has a high cation exchange capacity [23,52] and would be expected to have high nutrient retention capacity. Although the application of bentonite for silviculture has not been widely reported [72], we would expect its use to be similar to normal agriculture [37].

Corncob was inferior in terms of increasing water content as compared to charcoal and bentonite. With the addition of corncob, teak seedlings did not show growth acceleration in the first half of the experimental period, but growth recovered in the second half. Thus, corncob may provide positive effects on the growth of teak after its decomposition. For application in silviculture, the composting of corncob would be essential. Corncob had the capacity to suppress the uptake of Al, and this might be beneficial for silviculture in acidic environments.

ACKNOWLEDGMENTS

We would like to thank researchers of Royal Forest Department (RFD), Masato Oda, and Takuya Kajimoto for their invaluable comments on this study. We are grateful to the technical staff of Northeast Forest Seed Center, Khon Kaen for their technical assistance. Thanks are due to the technical staff of Plantation Silviculture Sub-division and Forest Soil sub-division for their support with our experiment. For the transportation of samples and analysis of plant organs, we also thank the staff of Japan International Research Center for Agricultural Science (JIRCAS). This research was conducted as a joint research project between the RFD and JIRCAS as part of a program for the improvement of techniques for the utilization of forest resources to promote sustainable forestry.

AUTHOR CONTRIBUTIONS

M.K. and R.Y. conceived the experiments. S.N. raised teak seedlings at the Seed Center in Thailand. M.K. performed the experiments, measured growth, photosynthetic rates and water potential, and analyzed various nutrients. S.H. supported measurements of M.K. in Thailand. All authors discussed the results, and co-wrote the paper.

REFERENCES

1. Tewari, D.N. *A Monograph of Teak (Tectona Grandis Linn. f.)*; International Book Distributors: Dehra Dun, India, 1992; p. 479.

2. Kaosa-ard, A. Teak (*Tectona grandis* Linn. f.) its natural distribution and related factors. *Nat. Hist. Bull. Siam Soc.* **1989**,*29*, 55–74.

3. Tanaka, N.; Hamazaki, T.; Vacharangkura, T. Distribution, growth and site requirements of teak. *Jpn. Agric. Res. Q.***1998**, *32*, 65–77.

4. Yoshifuji, N.; Kumagai, T.; Tanaka, K.; Tanaka, N.; Komatsu, H.; Suzuki, M.; Tantasirin, C. Inter-annual variation in growing season growth of a tropical seasonal forest in northern Thailand. *For. Ecol. Manag.* **2006**, *229*, 333–339.

5. Priya, P.B.; Bhat, K.M. Influence of rainfall, irrigation and age on the growth periodicity and wood structure in teak (*Tectona grandis*). *IAWA J.* **1999**, *20*, 181–192.

6. Leroy, C.; Saint-André, L.; Auclair, D. Practical methods for non-destructive measurement of tree leaf area. *Agroforest. Syst.* **2007**, *71*, 99–108.

7. Grace, J.; Fasehun, F.E.; Dixon, M. Boundary layer conductance of the leaves of some tropical timber trees. *Plant Cell Environ.* **1980**, *3*, 443–450.

8. Cernusak, L.A.; Winter, K.; Aranda, J.; Virgo, A. Transpiration efficiency over an annual cycle, leaf gas exchange and wood carbon isotope ratio of three tropical tree species. *Tree Physiol.* **2009**, *29*, 1153–1161.

9. Kunert, N.; Schwendenmann, L.; Hölscher, D. Seasonal dynamics of tree sap flux and water use in nine species in Panamanian forest plantations. *Agric. For. Meteorol.* **2010**, *150*, 411–419.

10. Rao, P.B.; Kaur, A.; Tewari, A. Drought resistance in seedlings of five important tree species in Terai region of Uttarakhand. *Trop. Ecol.* **2008**, *49*, 43–52.

11. Singh, K.P.; Srivastava, S.K. Seasonal variation in the biomass and non-structural carbohydrate content of fine roots of teak (*Tectona grandis* L. f.) plantations in a dry tropical region. *Tree Physiol.* **1986**, *1*, 31–36.

12. Calder, I.R.; Rosier, T.W.; Prasanna, K.T.; Parameswarappa, S. Eucalyptus water use greater than rainfall input—A possible explanation from southern India. *Hydrol. Earth Syst. Sci.* **1997**, *1*, 249–256.

13. Watanabe, Y.; Owusu-Sekyere, E.; Matsunaga, T.; Buri, M.M.; Oladele, O.I.; Wakatsuki, T. Teak (*Tectona grandis*) growth as influenced by soil physicochemical properties and other site conditions in Ashanti region, Ghana. *J. Food Agric. Environ.* **2010**, *8*, 1040–1045.

14. Zech, W.; Drechsel, P. Relationships between growth, mineral nutrition and site factors of teak (*Tectona grandis*) plantations in the rainforest zone of Liberia. *For. Ecol. Manag.* **1991**, *41*, 221–235.

15. Food and Agriculture Organization. *An Explanatory Note on the FAO World Soil Resources Map at 1:25,000,000 Scale*; World Soil Resources Reports 66; FAO: Rome, Italy, 1993; p. 64.

16. Kyuma, K. Soil resources and land use in tropical Asia. *Pedosphere* **2003**, *13*, 49–57.

17. Eswaran, H.; Padmanabhan, E. Tropics and sub-tropics, soils of the. In *Encyclopedia of Soil Science*; Lal, R., Ed.; Marcel Dekker Inc.: New York, NY, USA, 2002; pp. 1342–1345.

18. Shaw, J.N. Ultisols. In *Encyclopedia of Soil Science*; Lal, R., Ed.; Marcel Dekker Inc.: New York, NY, USA, 2002; pp. 1357–1361.

19. Gopikumar, K.; Varghese, V. Sand culture studies of teak (*Tectona grandis*) in relation to nutritional deficiency symptoms, growth and vigour. *J. Trop. For. Sci.* **2004**, *16*, 46–61.

20. Barroso, D.G.; Figueiredo, F.A.M.M.A.; Pereira, R.C.; Mendonça, A.V.R.; Silva, L.C. Macronutrient deficiency diagnosis in teak seedlings. *Rev. Árvore* **2005**, *29*, 671–679, (In Portuguese and English Summary).

21. Zhou, Z.; Liang, K.; Xu, D.; Zhang, Y.; Huang, G.; Ma, H. Effects of calcium, boron and nitrogen fertilization on the growth of teak

(*Tectona grandis*) seedlings and chemical property of acidic soil substrate. *New For.* **2012**, *43*, 231–243.

22. Salifu, K.F. Site variables controlling teak (*Tectona grandis*) growth in the high forest zone of Ghana. *J. Trop. For. Sci.***2001**, *13*, 99–108.

23. Tangmitcharoen, S.; Nimpila, S.; Phuangjumpee, P.; Piananurak, P. Two-year results of a clonal test of teak (*Tectona grandis* L. f.) in the northeast Thailand. In *Approach to Sustainable Forestry of Indigenous Tree Species in Northeast Thailand*; JIRCAS Working report 74; Noda, I., Vacharangkura, T., Himmapan, W., Eds.; Japan International Research Center for Agricultural Sciences: Tsukuba, Japan, 2012; pp. 19–22.

24. Suzuki, S.; Noble, A.D.; Ruaysoongnern, S.; Chinabut, N. Improvement in water-holding capacity and structural stability of a sandy soil in northeast Thailand. *Arid Land Res. Manag.* **2007**, *21*, 37–49.

25. Lesturgez, G.; Poss, R.; Noble, A.; Grünberger, O.; Chintachao, W.; Tessier, D. Soil acidification without pH drop under intensive cropping systems in northeast Thailand. *Agric. Ecosyst. Environ.* **2006**, *114*, 239–248.

26. Abod, S.A.; Siddiqui, M.T. Fertilizer requirements of newly planted teak (*Tectona grandis* L. f.) seedlings. *Pertanika J. Agric. Sci.* **2002**, *25*, 121–129.

27. Wichiennopparat, W.; Wanpinit, M.; Nimpila, S. A preliminary result of soil improvement trial of teak in Khon Kaen, Thailand. In *Approach to Sustainable Forestry of Indigenous Tree Species in Northeast Thailand*; JIRCAS Working report 74; Noda, I., Vacharangkura, T., Himmapan, W., Eds.; Japan International Research Center for Agricultural Sciences: Tsukuba, Japan, 2012; pp. 23–26.

28. Ramesh, V.; Korwar, G.R.; Mandal, U.K.; Prasad, J.V.N.S.; Sharma, K.L.; Yezzu, S.R.; Kandula, V. Influence of fly ash mixture on early tree growth and physicochemical properties of soil in semi-arid tropical alfisols. *Agroforst. Syst.* **2008**,*73*, 13–22.

29. Mishra, P.K.; Prasad, S.S.; Babu, B.M.; Varalakshmi, L.R. Bentonite as an ameliorant in an alfisol–a laboratory study. *J. Irrig. Drain. Eng.* **2001**, *127*, 118–122.

30. Kammann, C.I.; Linsel, S.; Gößling, J.W.; Koyro, H.W. Influence of biochar on drought tolerance of Chenopodium quinoa Willd and on soil-plant relations. *Plant Soil* **2011**, *345*, 195–210.

31. Novak, J.M.; Busscher, W.J.; Watts, D.W.; Amonette, J.E.; Ippolito, J.A.; Lima, I.M.; Gaskin, J.; Das, K.C.; Steiner, C.; Ahmedna, M.; Rehrah, D.; Schomberg, H. Biochars impact on soil-moisture storage in an ultisol and two aridisols. *Soil Sci.* **2012**, *177*, 310–320.

32. Kuan, Y.H.; Liong, M.T. Chemical and physicochemical characterization of agrowaste fibrous materials and residues.*J. Agric. Food. Chem.* **2008**, *56*, 9252–9257.

33. Karanasios, E.; Papadi-Psyllou, A.; Karpouzas, D.G.; Tsiropoulos, N.G. Optimization of biomixture composition and water management for maximum pesticide dissipation in peat-free biobeds. *J. Environ. Qual.* **2012**, *41*, 1787–1795.

34. De Lucia, B.; Cristiano, G. Composted amendment affects soil quality and hedges performance in the Mediterranean urban landscape. *Compost Sci. Util.* **2015**, *23*, 48–57.

35. Himmapan, W.; Noda, I.; Furuya, N. The study on the administration of private forest plantation cooperative of Thailand: A case study of Nomgbua Lamphu private forest plantation cooperative limited. *J. For. Manag.* **2010**, *4*, 1–12.

36. Wichiennopparat, W.; Wanpinit, M.; Visaratana, T.; Noda, I.; Sukchan, S.; Sasrisang, A. *Soil suitability Map for Teak Plantation in Chaiyaphum and Khon Kaen Provinces*; RFD-JIRCAS Joint Research Project: Bangkok, Thailand, 2015; p. 52. (In Thai)

37. Saleth, R.M.; Inocencio, A.; Noble, A.; Ruaysoongnern, S. *Economic Gains of Improving Soil Fertility and Water Holding Capacity with Clay Application*; IWMI Research Report 130; International Water Management Institute: Colombo, Sri Lanka, 2009; p. 30.

38. Chung, R.; Wu, S. Effect of corncob compost on plant growth in an acid red soil. *Commun. Soil. Sci. Plant Anal.* **1997**, *28*, 673–683.

39. Van Reeuwijk, L.P. *Procedures for Soil Analysis*, 6th ed.; International Soil Reference and Information Centre: Wagningen, The Netherlands, 2002; p. 100.

40. Klute, A. *Methods of Soil Analysis, Part 1. Physical and Mineralogical Methods*, 2nd ed.; Soil Science Society of America Inc.: Madison, WI, USA, 1986; p. 1188.

41. Sparks, D.L.; Page, A.L.; Helmke, P.A.; Loeppert, R.H.; Soltanpour, P.N.; Tabatabai, M.A.; Johnson, C.T.; Sumner, M.E.*Methods of Soil Analysis, Part 3. Chemical Methods*; Soil Science Society of America Inc.: Madison, WI, USA, 1996; p. 1390.

42. American Public Health Association; American Water Works Association; Water Environment Federation. *Standard Methods for the Examination of Water and Wastewater*, 20th ed.; American Public Health Association: Washington, DC, USA, 1998; p. 1220.

43. Tondjo, K.; Brancheriau, L.; Sabatier, S.A.; Kokutse, A.D.; Akossou, A.; Kokou, K.; Fourcaud, T. Non-destructive measurement of leaf area and dry biomass in *Tectona grandis*. *Trees* **2015**, *29*, 1625–1631.

44. Larcher, W. *Physiological Plant Ecology*, 4th ed.; Springer: Berlin, Germany, 2003; p. 513.

45. Evans, J.R. Photosynthesis and nitrogen relationships in leaves of C_3 plants. *Oecologia* **1989**, *78*, 9–19.

46. Kayama, M.; Kitaoka, S.; Wang, W.; Choi, D.S.; Koike, T. Needle longevity, photosynthetic rate and nitrogen concentration of eight spruce taxa planted in northern Japan. *Tree Physiol.* **2007**, *27*, 1585–1593.

47. Enríquez, S.; Duarte, C.M.; Sand-Jensen, K.; Nielsen, S.L. Broad-scale comparison of photosynthetic rate across phototrophic organisms. *Oecologia* **1996**, *108*, 197–206.

48. Gratani, L.; Pesoli, P.; Crescente, M.F. Relationship between photosynthetic activity and chlorophyll content in an isolated *Quercus ilex* L. tree during the year. *Photosynthetica* **1998**, *35*, 445–451.

49. Niinemets, Ü. Components of leaf dry mass per area -thickness and density- alter leaf photosynthetic capacity in reverse directions in woody plants. *New Phytol.* **1999**, *144*, 35–47.

50. Terashima, I.; Miyazawa, S.; Hanba, Y.T. Why are sun leaves thicker than shade leaves? –consideration based on analyses of CO_2 diffusion in leaf. *J. Plant Res.* **2001**, *114*, 93–105.

51. Barnes, J.D.; Balaguer, L.; Manrique, E.; Elvira, S.; Davison, A. A reappraisal of the use of DMSO for the extraction and determination of chlorophylls a and b in lichens and higher plants. *Environ. Exp. Bot.* **1992**, *32*, 85–100.

52. Goto, S. Digestion method. In *Manual of Plant Nutrition*; Editorial Committee of Methods for Experiments in Plant Nutrition, Ed.; Hakuyusha: Tokyo, Japan, 1990; pp. 125–128. (In Japanese)

53. Horpibulsuk, S.; Yangsukkaseam, N.; Chinkulkijniwat, A.; Du, Y.J. Compressibility and permeability of Bangkok clay compared with kaolinite and bentonite. *Appl. Clay Sci.* **2011**, *52*, 150–159.

54. Chen, Y.; Shinogi, Y.; Taira, M. Influence of biochar use on sugarcane growth, soil parameters, and groundwater quality. *Aust. J. Soil Res.* **2010**, *48*, 526–530.

55. Erhart, E.; Hartl, W. Compost use in organic farming. In *Genetic Engineering, Biofertilisation, Soil Quality and Organic Farming*; Lichtfouse, E., Ed.; Springer: Dordrecht, The Netherlands, 2010; pp. 311–345.

56. Durga, V.V.K.; Gusta, S. Effect of vesicular arbuscular mycorrhizae on the growth and mineral nutrition of teak (*Tectona grandis* Linn. F.). *Ind. For.* **1995**, *121*, 518–527.

57. Gurumurthy, S.B.; Sreenivasa, M.N. Screening and selection of efficient VA mycorrhizal fungus for teak (*Tectona grandis* L.). *Kamataka J. Agric. Sci.* **1998**, *11*, 956–960.

58. Warnock, D.D.; Lehmann, J.; Kuyper, T.W.; Rillig, M.C. Mycorrhizal responses to biochar in soil—Concepts and mechanisms. *Plant Soil* **2007**, *300*, 9–20.

59. Lehmann, J.; Rillig, M.C.; Thies, J.; Masiello, C.A.; Hockaday, W.C.; Crowley, D. Biochar effects on soil biota—A review. *Soil Biol. Biochem.* **2011**, *43*, 1812–1836.

60. Wardle, D.A.; Zackrisson, O.; Nilsson, M.C. The charcoal effect in boreal forests: Mechanisms and ecological consequences. *Oecologia* **1998**, *115*, 419–426.

61. Makoto, K.; Tamai, Y.; Kim, Y.S.; Koike, T. Buried charcoal layer and ectomycorrhizae cooperatively promote the growth of *Larix gmelinii* seedlings. *Plant Soil* **2010**, *327*, 143–152.

62. Telfer, A.; Barber, J. Elucidating the molecular mechanisms of photoinhibition by studying isolated photosystem II reaction centers. In *Photoinhibition of Photosynthesis*; Baker, N.R., Bowyer, J.R., Eds.; BIOS Scientific Publishers: Oxford, UK, 1994; pp. 25–49.

63. Yamazaki, J.; Ohashi, A.; Hashimoto, Y.; Negishi, E.; Kumagai, S.; Kubo, T.; Oikawa, T.; Maruta, E.; Kamimura, Y. Effects of high light and low temperature during harsh winter on needle photodamage of *Abies mariesii* growing at the forest limit on Mt. Norikura in central Japan. *Plant Sci.* **2003**, *165*, 257–264.

64. Sanches, M.C.; Ribeiro, S.P.; Dalvi, V.C.; da Silva Junior, M.B.; de Sousa, H.C.; de Lemos-Filho, J.P. Differential leaf traits of a neotropical tree *Cariniana legalis* (Mart.) Kuntze (Lecythidaceae): Comparing saplings and emergent trees.*Trees* **2010**, *24*, 79–88.

65. Jayasankar, S.; Sudhakara, K.; Babu, L.C. Provenance variation in growth physiology, anatomical characteristics and foliar nutrient status of teak (*Tectona grandis*) seedlings. *J. Trop. For. Sci.* **2003**, *15*, 37–50.

66. Lawlor, D.W. *Photosynthesis*, 3rd ed.; BIOS Scientific Publishers: Oxford, UK, 2001; p. 386.

67. Radin, J.W.; Ackerson, R.C. Water relations of cotton plants under nitrogen deficiency. III. Stomatal conductance, photosynthesis, and abscisic acid accumulation during drought. *Plant Physiol.* **1981**, *67*, 115–119.

68. Hatta, H.; Darnaedi, D. *Phenology and Growth Habits of Tropical Trees*; National Science Museum: Tsukuba, Japan, 2005; p. 439.

69. Kikuzawa, K.; Lechowicz, M.J. *Ecology of Leaf Longevity*; Springer: Tokyo, Japan, 2011; p. 147.

70. Mai, J.; Herbette, S.; Vandame, M.; Cavaloc, E.; Julien, J.; Ameglio, T.; Roeckel-Drevet, P. Contrasting strategies to cope with chilling stress among clones of a tropical tree, *Hevea brasiliensis*. *Tree Physiol.* **2010**, *30*, 1391–1402.

71. Tigabu, S.M.; Savadogo, P.; Odén, P.C. Facilitation of forest landscape restoration on abandoned swidden fallows in Laos using mixing species planting and biochar application. *Silva Fenn.* **2012**, *46*, 39–51.

72. Brockley, R.P. Effects of different sources and rates of Sulphur on the growth and foliar nutrition of nitrogen-fertilized lodgepole pine. *Can. J. For. Res.* **2004**, *34*, 728–743.

Index

S

T

V